Enhancing participation and governance in water resources management

Enhancing participation and governance in water resources management: Conventional approaches and information technology

Edited by Libor Jansky and Juha I. Uitto

TOKYO · NEW YORK · PARIS

© United Nations University, 2005

The views expressed in this publication are those of the authors and do not necessarily reflect the views of the United Nations University.

United Nations University Press
United Nations University, 53-70, Jingumae 5-chome,
Shibuya-ku, Tokyo, 150-8925, Japan
Tel: +81-3-3499-2811 Fax: +81-3-3406-7345
E-mail: sales@hq.unu.edu general enquiries: press@hq.unu.edu
http://www.unu.edu

United Nations University Office at the United Nations, New York
2 United Nations Plaza, Room DC2-2062, New York, NY 10017, USA
Tel: +1-212-963-6387 Fax: +1-212-371-9454
E-mail: unuona@ony.unu.edu

United Nations University Press is the publishing division of the United Nations University.

Cover design by Rebecca S. Neimark, Twenty-Six Letters

Printed in the United States of America

ISBN 92-808-1120-7

Library of Congress Cataloging-in-Publication Data

Enhancing participation and governance in water resources management : conventional approaches and information technology / edited by Libor Jansky and Juha I. Uitto.
    p. cm.
 Includes bibliographical references and index.
 ISBN 9280811207 (pbk.)
 1. Water-supply—Management. 2. Water resources development.
3. Information technology. I. Jansky, Libor. II. Uitto, Juha I.
TD345.E54 2005
333.91—dc22   2005025292

Contents

List of figures.. vii

List of contributors.. ix

Introduction.. 1

1 Enhancing public participation and governance in water
 resources management.. 3
 Libor Jansky, Dann M. Sklarew and Juha I. Uitto

Part I: Conventional approaches 19

2 Public participation and water resources management: The
 case of West Sumatra 21
 Syafruddin Karimi

3 Public participation in the development of a management plan
 for an international river basin: The Okavango case............. 33
 Anthony R. Turton and Anton Earle

4 Transboundary environmental impact assessment as a tool for
 promoting public participation in international watercourse
 management 53
 Jessica Troell, Carl Bruch, Angela Cassar and Scott Schang

Part II: "Information technology" approaches ... 81

5 The Internet and e-inclusion: Promoting on-line public participation ... 83
  Hans van Ginkel and Brendan Barrett

6 Promoting public participation in international waters management: An agenda for peer-to-peer learning ... 98
  Dann M. Sklarew

7 Development of an email-based field data collection system for environmental assessment ... 120
  Srikantha Herath, Nguyen Hoa Binh, Venkatesh Raghavan, Hoang Minh Hien, Nguyen Dinh Hoa, Nguyen Truong Xuan

8 New directions in the development of decision support systems for water resources management ... 133
  Kazimierz A. Salewicz

Part III: Efforts by international organizations ... 155

9 Improving public involvement and governance for transboundary water systems: Process tools used by the Global Environment Facility ... 157
  Alfred M. Duda and Juha I. Uitto

10 Public participation and governance: A Mekong River basin perspective ... 180
  Prachoom Chomchai

Index ... 217

Figures

3.1	Map of perennial rivers in Africa and disputes over water ...	34
3.2	The Eastern National Water Carrier in Namibia	38
3.3	Participants at the Green Cross International Water for Peace Workshop in Maun, Botswana	41
3.4	Graphical representation of the First Generation Strategic Report on the Okavango River Basin...........................	43
3.5	Graphical representation of the Second Generation Strategic Report on the Okavango River Basin...........................	45
3.6	Graphical representation of the Third Generation Strategic Report on the Okavango River Basin, presented at the Third World Water Forum in Japan	47
5.1	Access to broadband in England and Wales, 2004	88
6.1	Public participation guidance from the EU Water Framework Directive	104
6.2	Key stages in a TDA/SAP process	110
7.1	System overview	124
7.2	Details of client side	124
7.3	Server-side details	125
7.4	Input data form for the user	126
7.5	The water infrastructure information system (WIIS) data structure	128
7.6	List of existing infrastructure with the inserted record on the central database	129

7.7	A sample list of database entries in the water infrastructure information system	130
8.1	Components of the decision-making process	134
8.2	Main building blocks of the decision support system	139
8.3	Map of the Ganges River basin	145
8.4	Example screen showing the selection of parameters for the strategic policy alternative	146
9.1	Growth of the GEF international waters portfolio since its establishment	161

Contributors

Brendan Barrett
Academic Programme Officer
Office of the Rector
United Nations University
53-70, Jingumae 5-chome
Shibuya-ku
Tokyo 150-8925
Japan

Nguyen Hoa Binh
PeaceSoft Solutions Co.
Hanoi
Viet Nam

Carl Bruch
Legal Officer
Division of Environmental Policy
　Implementation
United Nations Environment
　Programme – RONA
1707 H Street, NW, Suite 300
Washington, DC 20006
USA

Angela Cassar
Former visiting scholar
Environmental Law Institute
1616 P Street, NW, Suite 200
Washington, DC 20036
USA

Prachoom Chomchai
Professor Emeritus
Chulalongkorn University
Bangkok
Thailand

Alfred M. Duda
Lead Environmental Specialist
Global Environment Facility
1818 H Street, NW
Washington, DC 20433
USA

Anton Earle
Deputy Head
African Water Issues Research Unit
　(AWIRU)
Suite 17, Private Bag X1
Vlaeberg, 8018, Cape Town
South Africa

Hans J. A. van Ginkel
United Nations Under-Secretary-
　General and

Rector of the United Nations
   University
53-70, Jingumae 5-chome
Shibuya-ku
Tokyo 150-8925
Japan

Srikantha Herath
Senior Academic Programme Officer
Environment and Sustainable
   Development Programme
United Nations University
53-70, Jingumae 5-chome
Shibuya-ku
Tokyo 150-8925
Japan

Hoang Minh Hien
Department of Dyke Management and
   Flood and Storm Control
Hanoi
Viet Nam

Nguyen Dinh Hoa
Institute of Information Technology
   (ITI)
Vietnam National University
Hanoi
Viet Nam

Libor Jansky
Senior Academic Programme Officer
Environment and Sustainable
   Development Programme
United Nations University
53-70, Jingumae 5-chome
Shibuya-ku
Tokyo 150-8925
Japan

Syafruddin Karimi
Vice Dean for Academic Affairs
Faculty of Economics
Andalas University
Kampus Limau Manih
Padang 25163
Indonesia

Venkatesh Raghavan
Osaka City University
Osaka
Japan

Kazimierz A. Salewicz
System Analyst
Tamariskengasse 102/121
1220 Vienna
Austria

Scott Schang
Senior Attorney
Environmental Law Institute
1616 P Street, NW, Suite 200
Washington, DC 20036
USA

Dann M. Sklarew
The GEF International Waters
   Learning Exchange and Resource
   Network (IW:LEARN)
c/o IUCN USA
Director and Chief Technical Advisor
1630 Connecticut Avenue, NW,
   Suite 300
Washington, DC 20009-1053
USA

Jessica Troell
Staff Attorney
Environmental Law Institute
1616 P Street, NW, Suite 200
Washington, DC 20036
USA

Anthony R. Turton
Head
African Water Issues Research Unit
University of Pretoria
Pretoria
South Africa

Juha I. Uitto
Senior Evaluation Advisor
Evaluation Office
United Nations Development
   Programme

One UN Plaza (DC1-406)
New York, NY 10017
USA

Nguyen Truong Xuan
Center for Information Technology
  (CIT)

Hanoi University of Mining and
  Geology
Hanoi
Viet Nam

Introduction

1
Enhancing public participation and governance in water resources management

Libor Jansky, Dann M. Sklarew and Juha I. Uitto

Background

Water is essential to our survival. Nonetheless, over 1 billion people today cannot obtain enough clean water to meet their basic human needs (UNESCO-WWAP 2003). Water scarcity plagues 27 nations, and an additional 16 nations are considered water stressed (WRI 2004). The United Nations has also identified rising demand for water as one of four major factors that will threaten human and ecological health over the next generation (UNESCO-WWAP 2003). As public health, development, economy and nature suffer, ensuring access to clean water is rising towards the top of government agendas.

Governments throughout the world face common problems in addressing the growing water crisis. They struggle to manage water in ways that are efficient, equitable and environmentally sound. Improvements in water efficiency often demand significant capital investment and legal and economic reforms – means generally beyond the capacity of members of the public directly impacted by lack of clean water. Equitable allocation and stewardship of water resources also require detailed understanding of interrelated hydrodynamic, socio-economic and ecological systems. Such knowledge is often sorely lacking among those responsible for water decisions at the local, provincial and national scales (Chapters 3 and 7 in this volume).

Critical knowledge about water management is distributed across governments, non-governmental organizations (NGOs) and the water users themselves. Consider South Africans' constitutional right to clean water or Brazil's recent *Sede Zero* (Zero Thirst) programme (Barreto and da Silva e Luz, n.d.; South Africa 1996). Each initiative requires broad understanding: by regional water managers of basin-wide flows, by localities of gaps in rural and urban water availability, and by potentially affected users of resources from "hot spots" of degradation. To realize such water initiatives effectively, governments must solicit participation early on, then continue actively to involve numerous segments of their societies, including those most marginalized and most vulnerable to water limitation and impairment. A variety of tools are becoming available to support such efforts (see Chapters 3, 7 and 8 in this volume; Bruch et al. 2005). Regrettably, lack of awareness about these sorts of tools has severely limited their application. Hence, the world in 2005 remains far from systematic in its integration of public participation (P2) in water resources management decisions and their implementation.

The current volume aims to further global understanding of approaches and techniques for applying public participation to improve water resources management. As used herein, water resources management is the aggregate of policies and activities used to provide clean water to meet human needs across sectors and jurisdictions and to sustain the water-related ecological systems upon which we depend. In most circumstances, water management aims to address interests and integrate usage across hydrologically meaningful units, such as watersheds (all land that drains into a river, lake or aquifer, along with the body of water itself). Some management aspects, however, such as transboundary flows across multiple basins and inter-basin water transfers via channels or virtual water, may necessitate a broader geographical scope.

The International Association for Public Participation (IAP2) describes public participation as "any process that involves the public in problem solving or decision making and uses public input to make better decisions" (IAP2 n.d.). Note that this term differs slightly from "stakeholder involvement", which involves those affected by a decision as well as those able to influence its intended outcome (e.g. non-public stakeholders such as international donor agencies). By contrast, public participation aims actively to increase attention to and inclusion of the interests of those usually marginalized, e.g. politically disenfranchised minorities or poor people indirectly affected by water management (see Chapter 6 in this volume).

Principle 10 of the 1992 Rio Declaration on Environment and Development (UNCED 1992) emphasizes that environmental issues such as water management "are best handled with the participation of all concerned citizens". The Declaration urges nations to facilitate public participation through methods that increase (a) transparency, (b) participatory decision-making and (c) accountability. These elements may be described respectively as: (a) informing people of water management issues or activities that may affect them, (b) involving the public in decision-making regarding such activities, and (c) providing those adversely affected by these decisions and activities with the means for seeking redress. This volume will address all three of these elements in the context of water resources management.

Perspectives on applying public participation to water management

This book contains the papers presented at an international symposium organized in October 2003 by the United Nations University (UNU) and Tokyo University of Agriculture and Technology (TUAT). The objective of the symposium was to identify successful mechanisms, approaches and practices for promoting public involvement in water resources management. The symposium also examined the conditions that facilitate or hinder public involvement, as well as contextual factors that may limit the transfer of experiences from one watershed to another. The works herein should be regarded as a survey of expert insights, though by no means a definitive treatise regarding this emerging field.

The text is organized into several parts. This chapter introduces a set of issues pertinent to public participation in water resources management, including a preview of the approaches addressed in subsequent chapters. Case studies in these later chapters focus on conventional approaches, information technology and IT-based approaches, and approaches involving international institutions. The book concludes with a perspective on how these various approaches might be systematically transferred and applied across a wide diversity of hydrological, socio-political and cultural contexts. As such, the final chapter serves also as the preliminary sketch for a future definitive framework for public participation applied to water resources (a substantial effort beyond the scope of the current volume).

This basic review and assessment of watershed management activities is intended to provide reliable information on lessons learned and existing gaps. Such information is much needed to justify investment in watershed management activities and to orient such activities to areas where they are most needed. The assessment and approaches were selected to respond to identified needs. The collection also takes into account the characteristics of different audiences involved in water resources management – within contexts of both national and transboundary watersheds.

Water resources management without public participation

People are often denied the right to participate in water management decisions and policies that concern them, with sometimes tragic results. For instance, large dams for water supply and irrigation have forcibly displaced tens of thousands, even millions of intended beneficiaries, across India, Mauritania, Brazil and many other places around the world. Numerous news reports also highlight how inadequate governance continues to allow industries to poison their neighbours through tainted water supplies and fish in China, Indonesia and elsewhere. Meanwhile, some governments have even intentionally used water policy to harm the disenfranchised, such as Iraq's years of draining the wetlands upon which its Marsh Arabs depended for millennia (Cohen 1997).

The brief examples above illustrate how inadequate public participation in governments' water management can result in tremendous social upheaval and the violation of the basic human rights of their citizens. How can public participation in water management have the opposite effect – providing for our basic human need for water and ensuring no thirst? Below we summarize three approaches explored by participants in the UNU symposium.

Conventional approaches

By conventional approaches to public participation, we simply mean approaches that utilize processes to inform, consult, involve, collaborate with and empower the public. These approaches may entail various levels of public participation from merely keeping the public informed to a stage where only actions that have been decided upon in a participatory manner will be carried out (see IAP2 2003). Opening Part I of this book, Syafruddin Karimi illustrates that, even with the best of intentions, governments face daunting challenges to include participatory management of water resources under conditions of limited institutional capacity and substantial spatio-temporal variability in water quantity. As communities (such as West Sumatra in Indonesia) make the transition from small traditional social structures to large, market-based societies, the collaborative management of water is further complicated by the emergence of competing water usage interests across sectors and regions.

Anthony Turton and Anton Earle's case study from southern Africa demonstrates how public participation in an international context can be equally daunting. Historical contention between international interest groups and riparian nations in the Okavango River basin was overcome, in part, by the recognition of a set of principles guiding transboundary water resources management. These include: national sovereignty in foreign policy, government decision makers' need for greater knowledge, and the role of scientific experts in informing water management decisions. This process also shows how scientists may perform a valuable service in developing viable options for watershed management.

Jessica Troell et al. present transboundary environmental impact assessment (TEIA) as an important methodology for encouraging public participation in water resources planning. TEIA involves testing the implications of various decision scenarios for the natural and human environment across political boundaries. As such, the timing and manner in which participatory mechanisms are incorporated into the EIA process can have a tremendous impact on the ultimate utility of the resulting water management regime. Still, participation in TEIA remains in its early stages, with much to glean from other domains of subnational and international environmental planning.

The chapters in Part I suggest that public participation processes for water governance are often practised in isolation from one another. To date, there has been little transfer of public participation practices or lessons between water management initiatives. As a result, there remain many potentially useful, though underutilized, tools (e.g. TEIA, consensus-building, joint fact-finding, visioning). Furthermore, it is often unclear how to measure and determine the success of conventional public participation initiatives.

Information technology approaches

How information technology (IT) can be utilized to promote public participation in the management of water resources is the main question tackled in Part II of the book. IT contains a range of tools and technologies that can be used to enhance the process of public participation. They may include the Internet and its various applications, as well as non-networked decision support and geographical information systems. Hans van Ginkel and Brendan Barrett focus on the vast potential for IT to enhance public participation in decision-making. They note both IT's promise, in terms of greater inclusiveness and faster response to environmental stresses, as well as its potential challenges – such as the pursuit of IT as an end instead of a means to improve participation and effective water management. Other pertinent externalities and misconceptions are also presented.

As summarized by Dann Sklarew, public involvement initiatives should be based on a set of culturally and politically relevant principles. With respect to public participation in international waters management, in particular, the International Waters Learning Exchange and Resource Network (IW:LEARN) has established a collaborative platform for the international waters community, which is accessible on-line. IW:LEARN and its partners invite IW managers, interested members of the public and private sectors, and civil society at large to participate in the workshop series design, development and evaluation.

Srikantha Herath et al. specifically target the limited human and IT capacity of countries such as Viet Nam to collect adequate environmental data to make effective decisions. The authors have developed an Internet-based data collection and management system, built upon email and free open source software. The result is a semi-automated tool for transferring various sorts of data collected in the field – where email often is the most advanced IT available – to centralized Web servers for data management and processing. The tool has been designed to be flexible, thus easily adaptable to other environmental assessment activities in Viet Nam or elsewhere.

Kazimierz A. Salewicz outlines the recent developments in decision support systems (DSS) for use in water resources management. As DSS expands from desktop to the Internet, along with concurrent increases in IT processing and storage, there is great potential for applying DSS in real time to both long-term structural decisions (e.g. where to build a dam or channel) and short-term operational decisions (e.g. how much water to direct down various conduits). One of several challenges considered is how to meld intuitive user interfaces effectively with powerful databases and valid computational models.

These IT narratives collectively indicate an emerging "toolbox" for increasing public awareness and participation in water resources management. A common challenge will be to adapt and apply appropriate IT or ICT (information and communication technologies) tools to the specific needs of water resource managers in real time.

International approaches

As demonstrated in Part III of the book, international and regional organizations also play a vital role in enhancing public participation around transboundary water resources management. The goal is to extend public participation across political boundaries and to empower the broader public to participate in decision-making and monitoring relating to projects and actions that concern two or more countries. International financing of water management projects may catalyse the inclusion of participatory processes, as is the case with the policies of the Global Environment Facility (GEF). Alfred M. Duda and Juha I. Uitto focus on the participatory process tools that the GEF and its projects use to improve the management of transboundary waters. According to the GEF's policy, public involvement consists of three related, and often overlapping, processes: information dissemination, consultation and stakeholder participation. Projects across East Asian seas, the Black Sea/Danube basin and Lake Tanganyika have used information dissemination to transcend national and cultural boundaries in order to improve transboundary water management. Meanwhile, consultation with stakeholders in the Rio San Juan basin between Nicaragua and Costa Rica has involved over 100 stakeholder groups and partnerships at various scales.

Beyond specific water management projects, permanent international basin organizations also play a long-term role in promoting public participation to improve watershed management. Prachoom Chomchai echoes lessons from West Sumatra and Okavango in his examination of the challenges faced in transboundary water management within the Mekong River basin, one of the most densely populated and rapidly developing areas in the world. The Mekong provides a lifeline for over 57 million people across six countries (WRI 2003). Conflicting interests between civil society and modern development and between local and national priorities and contexts have escalated through both participatory and extra-participatory processes (e.g. protests). Thus, participatory processes alone are not sufficient, he argues. In addition, there is a need for institutional overhaul as well as rethinking of the process of international development in general. This conclusion also reaffirms findings from other basins (e.g. UPTW 2003).

Ending global thirst depends upon providing the public with a voice in water resource decisions that directly affect them. Where the public are not included in decisions that affect their welfare, the Mekong example shows that they may resist change, protest or otherwise obstruct implementation of such decisions. Donors such as the GEF may at times be crucial to the integration of public participation in water resources management across national boundaries. Nonetheless, identifying appropriate moments and mechanisms to involve the public in water resources management remains an ongoing challenge.

Those living along international watercourses, near international borders and far removed from central governments are particularly difficult to include in such decision-making. Yet, for the stakeholders, transboundary participation is also critical. Inadequate public participation in transboundary water management has been a historical factor contributing to strife between nations along the Danube River, the Senegal River, the Mekong River and Lake Chad, for instance. However, improving public participation across international boundaries also requires addressing difficult transboundary challenges (e.g. sovereign water rights; migratory populations; linguistic and cultural differences; and distinct political, economic and legal frameworks among riparian nations). Nonetheless, public involvement, associated with ongoing reform in governance, holds the promise of improving the management of watercourses and reducing the potential for national and international conflict over water issues (see Chapters 3, 9 and 10 in this volume).

To realize this potential will require a more comprehensive understanding and systematic application of public participation processes across both national and international basins. This should begin, first and foremost, with a review of existing approaches and the tools available. Such an effort began with several workshops in early 2003 (Bruch et al. 2005; UPTW 2003). The current volume builds upon these important precursors by focusing on several specific approaches. Moving forward, mechanisms are now needed to systemize the experiences captured in these volumes and elsewhere, then transfer such guidance to those fostering public participation in water resources management around the world (see Chapter 6 in this volume).

Outstanding issues

It is clear that much progress has been achieved in water resources management, as new approaches and methodologies have been developed to promote participation across national and transboundary watersheds. Participatory processes are recognized as important in watershed management from project identification through design and implementation to monitoring and evaluation. Similarly, attention must be given to the services and benefits that participatory watershed management can provide. Watershed management is increasingly seen as an appropriate vehicle not only for environmental conservation but also for the improvement of rural and urban communities' living conditions. In this regard, there is a demand for the development of appropriate technologies, including ICT, that can ensure the sustainable development and management of natural resources involving the public.

Capacity development for participatory watershed management is one of the most needed parts of watershed development projects. In this respect, it is being recognized that there is a need for improved understanding and identification of the institutional and organizational arrangements required for effective watershed management. An appropriate legislative framework to support watershed management policies needs particular attention.

Nevertheless, several issues of concern emerge from the papers included in this volume, as well as the associated discussions in the UNU symposium. We deal with some of these below.

There is a need to define more clearly and adapt key terms to promote public participation in water governance

The role of public participation

The Rio Declaration, the GEF and IAP2 each describe key elements of participatory processes. Turton and Earle have also provided a set of participation principles related in transboundary water management in Chapter 3. Is this set of characterizations sufficient to guide practitioners?

During the symposium, Norio Okada suggested that defining factors could also include: (1) the level of public participation; (2) the process of public participation, including who initiated the process and who participated at each stage; (3) the communication platform for public participation; (4) the role of facilitation and consultation; and (5) the role of science and technology. In addition, he proposed that resources management involving local communities takes place at five different levels: (1) life in the community; (2) land use/built environment; (3) infrastructure; (4) social schemes/culture and convention; and (5) the natural environment. Each of these five spheres has its own speed of functioning, and these different levels are connected vertically and not horizontally. This way of thinking is needed in dealing with resources management so that adaptive management can be used for implementing ideas in the real world.

These various characterizations should be scrutinized in adapting a definition of public participation that is pertinent to water resources management, in particular. The result should clearly describe how public participation fits into adaptive water management regimes at local, national and international watershed scales.

Geographical scale and focus

Whereas water resources management has historically used the river or lake basin as its unit of management, the growth in population and urbanization has placed increasing pressure on water resources across multiple basins in a region (e.g. southern Africa, India, Tigris–Euphrates area, Tokyo Bay). Increasing scale from basins with millions of people to multi-basin regions of hundreds of millions may strain the capacity of any known public participation process. However, there is historical precedent for coordinated or "nested" management of basins at various scales (e.g. the North American International Joint Commission, Turkish–Syrian water management agreements). These may point to a future, multi-tiered approach to involving the public in managing their water resources.

Although groundwater accounts for about half of current potable water supplies, 40 per cent of the demand of self-supplied industry and 20 per cent of water use in irrigated agriculture (UNESCO-WWAP 2003), this fact is largely neglected in water resources management discussions. It is also important to bear in mind that the dimensions of groundwater aquifers are often uncertain and frequently cross jurisdictional and international boundaries. This lack of knowledge further complicates raising awareness and promoting public involvement. Still, such participation is crucial to ensure that these often slowly recharging systems are neither permanently depleted nor polluted beyond further use by those who depend on groundwater.

Scope of involvement: The public or stakeholders?

On a case-by-case basis, it is equally important to clarify whether participation is aimed at the public-at-large or at more specific interest groups, including non-public stakeholders. Stakeholders are often seen as only the people living in specific project areas. However, successful water resources management should ideally from the outset consider all people and institutions directly or indirectly affected by the project (though, practically speaking, not all stakeholders can be consulted directly). These would thus include, for example, sectoral ministries and government agencies not directly involved in the project but affected by its results. Whether and how to involve stakeholders physically outside the area of affected water resources are fundamental questions whose answers vary from case to case. Without such clarity, however, it would be difficult to determine success.

It is equally important to learn from local people, to respect different attitudes and experiences, and to seek out win–win situations based on such learning. Even in the twenty-first century, there continues to be a strong need for respected experts and decision makers to solicit and accept the different viewpoints of the affected public. These public stakeholders are, in fact, experts in their own right on the quantity, quality, usage and habitat associated with their personal and community water resources. With finesse and some luck, integrating local interests – sometimes viewed as parochial or self-serving – with other broader interests can result in the discovery and pursuit of a better, shared vision and strategy for collective benefit.

Measures of successful public participation

Monitoring and evaluation indicators are frequently used to measure the progress and impact of water resources management activities. Such indicators of success are also needed to track the success of public participation as it contributes both to water management as well as to broader societal goals, such as good governance. Specific public participation indicators should measure both progress (e.g. the development of and timely adherence to a stakeholder involvement plan, broad acceptance of a collective "watershed vision", the creation of basin-wide citizen advisory committees, etc.) and its impact (e.g. the public are generally satisfied with the result, or indicate being better off thereafter). Duda and Uitto have shown in Chapter 9 how transparent monitoring of indicators can also enhance public participation, increase the accountability of authorities and lead to a better performance and compliance with agreed norms.

As The Access Initiative (TAI) notes, "transparent, participatory, and accountable governance [is] an essential foundation for sustainable development" (TAI n.d.). TAI's standardized qualitative approach to tracking governments' progress towards realizing Rio Principle 10 could be adapted and applied to the water resources management domain. When linked to a clear vision or description of success, ongoing measures of such indicators will be key to determining the overall success of the public participation process.

There is a need for greater focus on institutional issues to ensure effective water resources management

Different countries are at different stages of development. Some countries are still struggling with basic infrastructure, whereas others have progressed to a need for greater institutional reform and capacity development. Within the span of countries at different stages of development, there may be valuable opportunities to work together and to exchange experiences and lessons in order to solve problems related to their water resources. Institutional development is especially important in terms of land tenure, economic, legal and policy reforms. All are crucial for effective community-level participation in water resources management.

During the symposium, Dann Sklarew stressed the need for transboundary legal frameworks and institutions to determine when and how to integrate public participation into design and implementation. How can momentum be maintained despite changes over time in the role of public participation as well as the make-up of those involved in water resources management for a given basin or area?

Since each river basin is unique, it may be difficult to transfer experiences from one place to another. In the case of sub-Saharan Africa, in particular, such transference is compounded by the fact that many countries are still proceeding from conflict situations to an absence of hostility and to ongoing peace. For instance, although the GEF Lake Tanganyika project was developed concurrent with the Rwandan genocide, it was only after violence had subsided a bit that riparian states were able to sign the resulting Convention. In this and other regions, an evolution in national governance and international relations is a crucial prerequisite for effective water resources management.

Public participation may provide the means for making the transition from dependence to empowerment

Based on her experiences on the Rio San Juan in Costa Rica, Hiromi Yamaguchi drew attention during the symposium discussion to how the term "public participation" is often viewed sceptically in developing countries. This phenomenon is also cited by Anthony Turton here. In some instances, local people and government officials may cynically consider "public participation" to be a key to getting more funds from donor agencies.

Experience also shows that foreign aid from donors can lead to a culture of dependence on the part of recipient countries. Fortunately, developing formal processes for public awareness and participation can be an effective means to increase local ownership, thereby laying the basis for locally sustained stewardship of water resources. Moreover, active oversight by donors and civil society may also ensure that water resources management projects and basin organizations are not spoiled or co-opted by corruption.

Determining appropriate roles for new technologies

Naruemon Pinniam Chanapaithoon emphasized during the symposium discussion that most or all new knowledge and technologies search for the same solution: how to allocate limited resources to all inhabitants in a fair and sustainable manner. This is the case for any resource, be it fresh water, food, clean air or fuel. The trick, as van Ginkel and Barrett emphasize in Chapter 5, is that new technologies are applied to meet such needs with conscientious attention to the externalities of such application.

Environmental impact assessment

Environmental impact assessment (EIA) has become an important and widely used tool. It is essential that EIA be used appropriately, not just as a rubber stamp for development projects. There have been many cases where EIA has been the only review mechanism applied and at only one point in time during project preparation. Once the EIA was completed, the project proponents have felt that they have fulfilled the review requirements and proceeded with the project. Those affected by the process are henceforth ignored. As a result, many people in developing countries see EIA as a negative governmental tool. Public perception and ongoing participation are very important to long-term success. EIA must become a continuous process that involves the local stakeholders who will be affected by the project in the overall, iterative process of water resources management.

Furthermore, in basins that span national or international jurisdictions or that cover multiple basins owing to long-distance water transfers, as mentioned above, tools such as the transboundary environmental impact assessment must be further developed and applied consistently across jurisdictions.

Information technology

We live in an era of information technology, and those who can access more information and know how to utilize it and make benefit of it wield increasing power. Therefore, what is crucial today is not only a fair distribution of resources but also equal chances for people to acquire information to use those resources appropriately. Although the Internet provides a powerful medium for collecting and disseminating information, over-reliance on computers for public awareness may limit participation to those on the "virtual" side of the so-called "digital divide". It is thus important that the Internet be leveraged to communicate with stakeholders beyond the computer *literati*, including people without any direct access to IT.

The public have a right to know the assumptions, costs and benefits of water resource decisions that affect them, through government transparency and accountability. In order to become useful information, data should be unbiased and undistorted before reaching their audience. In developing countries in particular, quality – as well as quantity – of data issues becomes critical. Such quality does not always come cheaply. Getting good enough data sufficient for decision-making in developing countries is often difficult, so one must be very careful about the kinds of conclusion that can be drawn and the decisions to be made. On the other hand, the choice is often between using inadequate data or no data at all. In such cases, the public must be informed (in understandable language) of the limitations of the data available and the resulting assumptions used by decision makers.

Geographical information systems

Geographical information systems (GIS) are recognized as very powerful tools for decision-making. GIS can be used to examine different problems, proposals, ideas, etc., in the quest for effective water management decisions. The quality of data is very important because that affects the answers that we get to the questions posed. It is important to bear in mind that GIS and other information technologies are only tools, and there is a risk of their oversimplifying the problem-solving process, i.e. assuming that just applying GIS technologies will solve the problems.

GIS and other information technology tools can be powerful in providing an improved basis for decision-making, but they cannot offer a panacea. It is essential to understand that the choice of actions to be followed must be based on a careful analysis of the social, cultural, political and economic factors, and that the choice is ultimately political in nature.

Developing countries often do not have sufficient financial resources to make use of sophisticated technologies. In fact, funds are often too limited to use the modern technologies and expertise common in more prosperous countries. It is thus important that the tools alone are not allowed to determine the agenda. Information technology should be used appropriately for different cultural contexts and available media, rather than trying to use sophisticated tools where they cannot be operational.

After examining various approaches, water resource managers must then select and integrate

Traditional and information technology approaches are both useful and complementary in promoting improved public participation in water resources development. Traditionally, access to information that enables decision-making has been limited to very narrow lines of professionals and administrators. Today, thanks to technological development, information has been made more accessible to a wider range of people (although more is obviously still needed). Now the question is how to use that information properly.

Government bureaucrats and policy makers for a given body of water may change over time. When talking about global water resources, however, stakeholders are not only the people in a certain region; all of us around the world are affected. Even with the help of various decision support systems to cope with a huge amount of complicated data, water policies and plans cannot be implemented successfully without the cooperation of the public as a whole. The tools cannot give the whole correct answer if the input is wrong from the beginning, if the analysts cannot distinguish the cause and the impact of what they are doing, or if the analysis does not lead to any reliable alternatives useful for the society. Therefore, learning how to use a tool is only one step in a process that must include how to apply it effectively and make use of it properly.

Similarly, we must continue to learn and transfer successful approaches to involving the public in water management decisions that affect them at local, national, regional and global scales. As populations grow and migrate over time, such broad inclusiveness is critical to early warning and rapid response to emerging water concerns. Only then will we collectively be able to manage Earth's water efficiently, equitably and ecologically and at the same time ensure clean water for all humanity.

It is recognized that significant progress on participatory watershed management approaches and methodologies has been achieved in different places in the world. An important challenge will be the dissemination and exchange of information about achieved success and lessons learned between institutions within the same country and across basins involving one or more countries.

REFERENCES

Barreto, A. N. and M. J. da Silva e Luz (n.d.) *Sede Zero – Um Desafio Hídrico para o 3.° Milênio* [Zero Thirst – A Water Challenge for the 3rd Millenium], ⟨http://www.ambientebrasil.com.br/composer.php3?base=./agua/doce/index. html&conteudo=./agua/doce/artigos/sedezero.html⟩ (accessed 17 June 2005).

Bruch, C., L. Jansky, M. Nakayama and K. A. Salewicz, eds (2005) *Public Participation in the Governance of International Freshwater Resources.* Tokyo: United Nations University Press.

Cohen, R. (1997) "ICE Case Studies: Marsh Arabs, Water Diversion, and Cultural Survival". Inventory of Conflict and Environment Case Study No. 60, ⟨http://www.american.edu/ted/ice/marsh.htm⟩ (accessed 17 June 2005).

IAP2 [International Association for Public Participation] (n.d.) *IAP2's Public Participation for Decision Makers,* ⟨http://iap2.org/training/decision-fact-sheets. pdf⟩ (accessed 17 June 2005).

────── (2003) *Foundations of Public Participation,* ⟨http://www.iap2.org/ indexpdfs/foundations-bro.pdf⟩ (accessed 17 June 2005).

South Africa (1996) *Constitution of the Republic of SA (Act No. 108 of 1996),* Section 27 (1) (b), ⟨http://www.info.gov.za/documents/constitution/1996/96cons2. htm⟩ (accessed 17 June 2005).

TAI [The Access Initiative] (n.d.) "About TAI", ⟨http://www.accessinitiative.org/ about.html⟩ (accessed 17 June 2005).

UNCED [United Nations Conference on Environment and Development] (1992) *Report of the United Nations Conference on Environment and Development (Rio de Janeiro, 3–14 June 1992). Annex I: Rio Declaration on Environment and Development.* UN Doc. A/CONF.151/26 (Vol. 1), ⟨http://www.un.org/ documents/ga/conf151/aconf15126-1annex1.htm⟩ (accessed 17 June 2005).

UNESCO-WWAP [United Nations Educational, Scientific and Cultural Organization World Water Assessment Programme] (2003) *Water for People, Water for Life. The United Nations World Water Development Report.* Barcelona: UNESCO and Berghahn Books.

UPTW [Universities Partnership for Transboundary Waters] (2003) "Stakeholder Participation in International River Basins: Models, Successes and Failures", Workshop. Corvallis, OR, 14–16 April, ⟨http://waterpartners.geo.orst.edu/new. html#past⟩ (accessed 17 June 2005).

WRI [World Resources Institute] (2003) *Water Resources eAtlas. Watersheds of the World. AS19 Mekong*, ⟨http://multimedia.wri.org/watersheds_2003/as21.html⟩ (accessed 17 June 2005).

——— (2004) *Water Resources and Freshwater Ecosystems – Actual Renewable Water Resources: Per Capita*, ⟨http://earthtrends.wri.org/searchable_db/index.cfm?theme=2&variable_ID=694&action=select_countries⟩ (accessed 17 June 2005).

Part I
Traditional approaches

2

Public participation and water resources management: The case of West Sumatra, Indonesia

Syafruddin Karimi

Introduction

Some may believe water to be the bounty of God, but its availability for human needs does not come without effort. Economic development and population growth have increased demand for a scarce economic resource whose supply does not grow in line with demand. Indeed, the increasing demand for water has even reduced its supply, which may lead to a water crisis.

Neither the market mechanism nor state intervention works adequately to match the water supply to the need for water. The potential water supply does not necessarily translate into an actual water supply that can readily meet actual demand. There is almost always a shortage of water supply. Faced with persisting water market disequilibrium, individual water users' behaviour leads to the "tragedy of the commons": in the absence of appropriate water resources management, the tendency is for individuals to protect their self-interest rather than the collective interest. Realizing the shortcomings of water resources management, Indonesia enacted a new water resources law in 2004. The law is expected to accommodate the role of public participation in water resources management and to overcome institutional as well as legal constraints on improving water resources management. In relation to the new water resources law, this chapter reports on the practice of public participation in water resources management in West Sumatra in Indonesia.

The tradition of public participation in West Sumatra

West Sumatra is a province of Indonesia on the island of Sumatra. The total area of West Sumatra is around 42,000 km^2. According to the Census of Population in 2000, the population of West Sumatra had reached 4.5 million. The population growth rate was less than 2 per cent and the population density per km^2 increased from 81 in 1980 to 100 in 2000. The population of West Sumatra is predominantly Minangkabau by tradition and is noted for its matrilineal family system.

The economy of West Sumatra is dominantly characterized by agriculture in terms of the working population: almost 50 per cent of the working population lives from agriculture. However, agriculture accounts for only around 30 per cent of the value of the regional domestic product of West Sumatra, whereas the manufacturing sector, which accounts for only 6 per cent of the working population, contributes more than 11 per cent to the regional domestic product. The economy grew by around 7 per cent per annum during 1966–2000. Real income per capita in 2000 was around Rp 5 million (roughly equivalent to US$600).

According to Minangkabau traditions, any public activity can be implemented as long as the planned activity has been decided through the process of *musyawarah* (public consultation) to reach *mufakat* (public consensus). Local community affairs always involve the public by means of this process. In principle, *musyawarah* for *mufakat* is the traditional Minangkabau value of public participation, as it is called today. Under the "New Order" government, *musyawarah pembangunan* (public consultation for development) was a forum for the government to control public preferences. Public participation was misused to mobilize local communities and manipulate their participation in the direction desired by the government.

The substance of public participation is actually rooted in the traditional values of the Minangkabau community. Before the New Order government, local communities were actively participatory and undertook their own development activities. They were involved in preparing, designing, planning, implementing and financing the process of development. Local communities were not used to depending on state intervention; their social capital was strong enough to generate the enthusiasm and energy needed for development. As the New Order government took over as the agent of development by centralizing the idea of development, local communities were left without a role. Central government's effect on local communities in West Sumatra was very significant when the Village Government Law imposed uniformity on all villages in Indonesia. All local rules had to conform to the centralized rules, and local community participation became pointless.

The Minangkabau people consider their traditional values to be a form of capital that enables them to design the future of community development. They tend to interpret the present social crises as a product of the political manipulation of their traditional values. According to their traditional values, the institutions of the Minangkabau community are based on an independent and autonomous *nagari* (autonomously democratic village government), which rules their social, political and economic affairs. Every *nagari* has its own local law to regulate activities within the domain of the *nagari*. It is understood by every *nagari* that no *nagari* will interfere in the business of another *nagari*. If there is an attempt, it will fail. Of course, there should be participation, but the development of participation should be through participatory methods that respect local norms. The norms of the Minangkabau community include a lot of terms that support participation. Everybody is always useful in the Minangkabau mindset. The blind, the deaf and the invalid have distinct functions in the system of public life in West Sumatra. Nobody is left out, and everybody counts. That is the traditional spirit of participation within the Minangkabau community of West Sumatra and it is the key to achieving public participation in West Sumatra.

Development efforts create benefits for a community. At the same time, they also require the community to sacrifice some resources in order to contribute to the costs of the development activities. The community rationally expects to receive more in benefits than it contributes to the cost. However, the reality does not always match expectations. On the one hand, some communities receive more in benefits than they contribute to the costs, or even do not contribute at all. On the other hand, there are communities that contribute more to the costs than they enjoy in benefits, or even do not benefit at all.

Problems always arise from the unequal distribution of benefits and costs, and they are not easy to solve through the centralized government alone. Solutions are even more difficult in the presence of a project-minded government. The success of often expensive development efforts depends on public participation; otherwise, the development is doomed to fail. Mere good intentions on the part of the government are not sufficient to guarantee the success of development efforts. Winning the heart of the community is often a necessary as well as a sufficient condition. This can be achieved by involving the community from the start.

We learnt a lot about success in the development of water resources for the community resulting from community participation. On the whole, the higher the level of community participation, the greater is the possibility that the project will end in success. We studied the development of water resources for the community in several villages of West

Sumatra. Public participation turns out to be the entry point for any government project for the betterment of the community (Karimi 2003).

When a government project for a community is finished, it is then expected that service delivery will function to the public benefit. However, there are cases where the finished projects do not function as expected. We found that the outcome was often worthless owing to conflicts between the community's aspirations and the government's self-serving decisions.

In practice, development activity was almost always related to the land area that belonged to the local community. It was common for the local community to ask for the land to be respected appropriately, whereas the government often asked the community to sacrifice the land for development. The government agreed to compensate the community only for valuable properties on land that would be useful for the development projects. The government often did not provide compensation for the land itself, which had the most value for the people. These issues of compensation appeared to exacerbate the problems of development projects with limited public participation.

An improved political climate after the collapse of the New Order government in 1998 opened the door for communities to express their aspirations and request protection for their rights. This encouraged public participation to develop. It is no longer possible for the government harshly to enforce its self-serving decisions in the name of the public interest. Involving the public is now becoming a prerequisite for achieving the development of public infrastructures and facilities. The government must consult the public in order to understand their preferences. Currently, the public are requesting that the government accommodate their aspirations, involvement and participation. This accommodation should not just be a paper exercise, but should pave the way to comprehensive public involvement. When the state loudly claimed to be an agent of development, there was also mention of public participation. But it was limited to warning the public to look after the construction produced by the development project. If there was public participation, it was merely for show. After three decades under the New Order government, there are a lot of lessons to be learned about improving public participation. The traditional values of the Minangkabau community in West Sumatra still exist to support the development of public participation.

Water resources management in West Sumatra

West Sumatra has four lakes and more than 600 rivers, and is categorized as a province with an abundance of water resources. The total water potential is estimated to be 41,965.51 million m^3, which comprises 41,944.40

million m³ of surface water and 21.13 million m³ of groundwater (Government of West Sumatra 2000). The total demand for water amounts to 6,745 million m³, which consists of 6,035 million m³ for irrigation and 709 million m³ for domestic consumption. The dominance of irrigation in water consumption is in line with the nature of West Sumatra's economic structure. This leaves a water balance of 34,346.5 million m³. Thus, total water demand utilizes only 16 per cent of the total availability of water.

Aside from the utilization of water resources for irrigating agricultural land, West Sumatra has made use of the Batang Agam River, Lake Maninjau and Lake Singkarak to power three hydroelectric plants, which are operated by PLN, the state-owned monopoly electricity company. These hydroelectric power stations contribute to the electricity supply for Sumatra as a whole.

The Singkarak hydroelectric power plant is the latest hydroelectric power station and its operation is reported to have reduced the flow of water into the Anai and Ombilin rivers, which has seriously disrupted the supply of water for agricultural activities in Solok and Tanah Datar regencies (Sudharta 2002). Before the Singkarak hydroelectric power plant came into operation, indigenous people from 13 *nagaris* (villages) in the region surrounding Lake Singkarak supported themselves by economic activities based on Lake Singkarak's water resources. The people mostly live from fishing in the lake, which is noted for its *bilih* fish. In a recent focus group discussion representing the indigenous people from around Lake Singkarak, it was learned that operation of the Singkarak hydroelectric power plant has reduced the local economic activities of both fishermen and farmers.

The supply of drinking water is an important challenge for water resources management in West Sumatra. Household demand for drinking water is met from various sources – bottled water, piped water, pumped groundwater, well water, spring water, river water and rainwater. Piped water is supplied only by PDAM (a local government body in charge of providing drinking water), whose market is mostly limited to urban areas that purchase water services. In 2001, PDAM produced 54 million m³ of clean drinking water (BPS 2001), up from 41 million m³ in 1995. This increased production also encouraged demand for piped water. In 1995, piped water as a source of drinking water accounted for 19 per cent of households (BPS 1996). This increased to 22 per cent in 2001 (BPS 2002).

Assuming that piped water is a safe standard for domestic consumption, the figures indicate that PDAM needs to expand its production and distribution networks in order to accommodate households that are not currently served. Otherwise, the majority of households will continue to fulfil their own needs for water. Pumping groundwater is the most important way for households to fulfil their needs for water. Although the proportion of households supplying their needs for water by pumping groundwater was only around 6 per cent, the trend has risen in recent years.

Public participation in water resources management

Baruah Bukik: Success in developing cooperation

A water resources development project for the local community in Baruah Bukik made full use of public participation. The project was funded by the Rural Water Supply and Sanitation (RWSS) project. Before the project, the local people of Baruah Bukik had to walk 250–750 metres to reach a source of water and they carried the water home in buckets made of bamboo. The RWSS project provided the materials to construct the infrastructure at the water source, and the local community participated by contributing their labour for free through the *gotong royong* (cooperative) system. The project successfully completed a water supply facility to meet the needs of a local community. Now that water is piped to every home, the locals have time to spare for other productive activities.

The success of the water project was the result of good cooperation between a government that understands the people's needs and a local community that has strong social capital in the form of *gotong royong*. This traditional social capital is not only alive within the local community of Baruah Bukik, but also on the increase. This strong social capital is enhanced by the local people who have mostly finished secondary school. They understand how to maintain their water supply and, by undertaking *musyawarah* (public consultation among the traditional community to reach consensus), they also established their own system for managing the water service for their community. The system comprises a Service Management Unit (SMU) and a Service User Group (SUG). In a local business meeting, the SMU and SUG together decided on the price of the water. Every home has to pay a deposit in advance of Rp 20,000 and a monthly fee of Rp 4,000. The monthly fee can be paid as Rp 1,500 in cash plus 1 litre of rice. As a fee-collecting unit, the SUG retains 30 per cent of the total fees collected to run its operation. The SMU, which receives the other 70 per cent, deposits its share in the People's Credit Bank (PCR). The SMU can use the money for the maintenance, operation and rehabilitation of the water facility.

Simabur: A failure to respect local rights

Following the Village Government Law of 1979, *nagari* Simabur was split into three villages: Simabur, Tanjung Limau and Koto Tuo. Of the three villages, Simabur is the capital, having adequate public facilities such as a market, an education centre, sources of public finance, and access to information and public transportation. Simabur is a hill village with a population of 2,097, more than half of whom live down in the valley. Public facilities, including water sources, are located in the valley area. Anyone living on the hill has to walking as much as 500 metres every day to fetch drinking water.

The local government of Tanah Datar wanted to help the local people of Simabur to have easy access to water. In 1995, PDAM installed a 3 km pipeline to channel water from Pincuran Tinggi (High Fountain) located in Sikaladi, a neighbouring village to Simabur. PDAM also constructed several public water tanks for local people to store water. After all the construction work had been completed, the people of Sikaladi village prohibited the use of the water source at Pincuran Tinggi. PDAM had failed to consult the indigenous local community of Sikaladi, which traditionally owns *ulayat* (communal) rights over the water source.

Sikaladi and Simabur belong to different *nagari*, which have different *adat* (traditional) rules. The local government should have positioned itself as a mediator between the people of Sikaladi, who own the traditional rights to the source of water, and the people of Simabur, who need water. In fact, the people of Sikaladi understand that they do not have absolute rights over the water source in Pincuran Tinggi. They believe that the water source actually belongs to God. The higher government needed skill in becoming involved in the indigenous community, whose local *adat* and *ulayat* rights deserve respect. In the case of Pincuran Tinggi, the government did not show this respect. Therefore, the government-initiated project had to accept failure.

Two years later, still being very eager to supply the need for water in Simabur, the government of Tanah Datar tried to utilize water sources located within *nagari* Simabur itself. The target was a water source in Bulakan at the mosque complex. However, PDAM saw the potential for profit if it could also connect the water source to Batu Sangkar, the capital of the Tanah Datar regency. Unfortunately, PDAM failed to share this profitable idea with the people, although they learned of PDAM's intentions soon enough. In response, the local community refused to allow PDAM to utilize the water resources in Bulakan. Again, the government had to accept failure, despite the fact that the water resource and the need for water were both located in the same village.

The government had made no attempt to consult the local community in order to understand its own priorities and preferences in utilizing water resources. The government had not learned from the failure in 1995. Public consultation for consensus is a necessary as well as a sufficient condition for achieving public participation by the traditional Minangkabau community. In the present era of democracy, public involvement, public consultation and public participation are a must for any government wishing to gain a good reputation. Otherwise, it will reap a bitter harvest of instability.

Public participation and water resources management in the new Water Resources Law

The strategy of water resources development and management in Indonesia has until recently emphasized the importance of meeting rapidly rising demand. This supply-side approach has led to overexploitation, which endangers the sustainability of water resources. The management of water resources is becoming merely the management of the water construction project.

In response to changing strategic conditions in water resources management, the government issued Law No. 7 in 2004 to regulate the management of water resources development. The new law states that water resources management should be based on the principles of sustainability, public benefit, integration and consistency, justice, autonomy, transparency and accountability. In order to achieve sustainable water resources with maximum public benefit, it is also emphasized that water resources should be managed in a comprehensive, integrated and environmentally friendly way, thus fulfilling the social, environmental and economic functions of water resources.

In order to protect the rights of every individual citizen to acquire a daily minimum basic water consumption to achieve a healthy, clean and productive life, water resources are controlled by the state and are utilized to maximize public prosperity. Central and local governments act for the state in controlling water resources. It is important to note that the state control of water resources admits the existence of *ulayat* rights traditionally owned by local communities. Based on its role of controlling water resources, the state assigns two types of water use rights, for individual consumption and for business operations. The state automatically allocates to every citizen water use rights for basic daily individual consumption and for agricultural activity within an irrigation system. The state requires citizens to get a permit for water consumption that changes the natural condition of water resources, for water volume above individual consumption and for agricultural activity beyond an irrigation system. Water use rights for business operations require a permit from the state. The state assigns water use rights to individual citizens as well as to private business institutions. The law states that these water use rights cannot be rented or transferred from the individual citizen or private business institution initially assigned the rights. In other words, the assigned water use rights are not marketable.

The new law also provides for public participation in water resources management. Specifically, the law provides the public with the rights to gain access to information related to water resources management, to receive appropriate compensation for losses caused by water resources management, to share in the benefits arising from water resources management, to express objections to announced plans for water resources management, to report complaints to the relevant authority regarding losses suffered owing to water resources management, and to bring to court claims related to problems caused by water resources management. Furthermore, the law provides the opportunity for the public to participate in planning, implementing and supervising the process of water resources management.

The law regulates public participation not only in terms of rights but also in terms of public obligations related to water use rights assigned by the state. In exercising water use rights, the public are obliged to participate in conserving water resources and in protecting and securing water resources infrastructure. Integrated action is necessary to balance the rights and the obligations in securing public participation. The law states the importance of integrated action among various stakeholders in water resources management to maintain the sustainable function and benefit of water and water resources. The state is authorized to establish a Water Resources Council as a coordinating mechanism to practise integrated water resources management at the national and local levels of government bodies.

In the event of conflict among stakeholders, the principle of *musyawarah* for *mufakat* is used as the main means of reaching resolution. However, the adoption of local traditional norms for resolving conflicts does not prevent the involved parties from bringing the case to court, as also provided for by the law.

Although it is mandatory for the state to make the supply of water available to everybody in the territory of Indonesia, it is the responsibility of central and local governments to carry out this commitment, subject to their capability. There is thus no guarantee that every citizen will gain access to water facilities as long as the government does not have the capability. According to the constitution, state control over water resources for the purpose of maximizing public prosperity tends to conflict with the nature of water as a scarce resource (Helmi 1997, 1998; Martius 1997). Therefore, using the market mechanism to determine water prices might act as a barrier to meeting basic human needs for water and sanitation, because the market mechanism does not discriminate between social functions and economic activities.

However, the new Water Resources Law does seem to promise a strong commitment to maximize water benefits for the public at large. The utilization of water resources must create a just and equitable improvement in people's standard of living. Access by every citizen to water for basic needs should increase as a result of improved utilization of water resources. The principle of justice must be clear to maintain the sustainability of water resources management. The ethics of sustainable water resources management needs to consider the redistribution aspect of water resources management so that every citizen can be freed from water poverty. The strong hold of the state over the water business is expected to improve the access of every citizen to water for basic needs. However, the state's responsibility to reduce water poverty for every citizen cannot be left to common responsibility as anticipated in the Water Resources Law. Otherwise, it would be better for the state to let the public take responsibility for securing their own basic water needs. Since common responsibility in practice means each citizen's responsibility, the result would be the maintenance of inequality of access to water and water poverty as a form of negative public goods.

The Water Resources Law places the state-owned enterprises as the main player in the development of the drinking water supply business, although other business players such as cooperatives, the private sector and local communities can also take part. Since the state is assuming the responsibility for meeting the needs of households for drinking water, the development of the drinking water supply system becomes the responsibility of the state as well. The state administers the development of the drinking water supply system in order to generate better management and services that provide drinking water at a price that the public are able to pay. At the same time, the state expects to maintain an equilibrium position between the consumers and the service providers.

The state has legally involved itself in the water business by using state-owned enterprises at the national and local levels. The business performance of state-owned enterprises makes it questionable whether they are capable of providing the public with access to water supplies, particularly in the era of globalization and privatization. Learning from the current trend of the government's privatization agenda, the state-owned enterprises are a stepping-stone to the privatization of all state-owned businesses (Center for Public Integrity 2003). The process of privatization is full of conflicting interests that mostly place a burden on public prosperity. In a democracy, public participation cannot be reduced to the role of parliament alone in making public decisions. Since the state is responsible for making water available to maximize public welfare, it is doubtful whether the state-owned enterprises, acting in the name of the state, will be able to fulfil the responsibility to provide public goods for public prosperity.

Conclusion

Public participation is necessary for water resources management. The implementation of a project with public participation creates public benefit without conflict, whereas a project without public participation creates public cost with conflict. Water resources management involves various stakeholders, and active public participation is necessary to integrate the potentially conflicting interests of these stakeholders, as is evident from the two water resources development projects in the Tanah Datar regency discussed above. The project at Baruah Bukik had a successful outcome in the presence of public participation, whereas the project at Simabur had to accept failure owing to the lack of public participation. Public participation thus reduces costs and increases productivity.

The new Water Resources Law in Indonesia has supported the need for more public participation in water resources management. The law recognizes the right to public participation as part of the ethics of water resources management. The law defines water use rights for individual consumption and for business operations; these rights are not marketable and are not transferable. The control of water resources belongs to the state, which assumes the responsibility for providing basic water needs. At the same time, the state-owned enterprises assume a dominant role in the water resources business to act as the agent of the state. However, an improvement in public participation and water resources management will depend upon a state commitment actually to implement the new law.

REFERENCES

BPS (1996) *Survei Sosial Ekonomi Nasional (SUSENAS) 1995*. Padang: Kantor Statistik Provinsi Sumatera Barat.
―――― (2001) *Statistik Air Bersih 2001*. Padang: Kantor Statistik Provinsi Sumatera Barat.
―――― (2002) *Survei Sosial Ekonomi Nasional (SUSENAS) 2001*. Padang: Kantor Statistik Provinsi Sumatera Barat.
Center for Public Integrity (2003) *The Water Barons: How a Few Powerful Companies Are Privatizing Your Water*. Washington, DC: Public Integrity Books.
Government of West Sumatra (2000) *A Profile of Water Resources in West Sumatra*. Padang: The Office of Irrigation.

Helmi (1997) "Ke Arah Pengelolaan Sumberdaya Air Yang Berkelanjutan: Tantangan dan Agenda Untuk Penyesuaian Kebijaksanaan dan Birokrasi Air di Masa Depan", *Visi Irigasi* 13(7).

────── (1998) "Memposisikan Status Air Sebagai Barang Ekonomi di Indonesia: Isu Konstitusi, Kebijakan dan Implementasi dalam Kerangka Memberikan Jaminan Air Bagi Petani", *Visi Irigasi* 14(7).

Karimi, S. (2003) "Community Participation toward Nagari Development", paper presented at the Workshop on "Infrastructure Development and Supra Structure for Improving Participation in the Nagari Community Economic Development", The Regency Government of Pariaman, Pariaman, 10 April.

Martius, H. (1997) "Penyesuaian Peran Birokrasi dan Pemberdayaan Ekonomi Petania: Etika Pendayagunaan Sumberdaya Air di Indonesia", *Visi Irigasi* 13(7).

Republic of Indonesia (2004) Water Resources Law, No. 7.

Sudharta, H. (2002) *Pembaharuan Kebijakan Pengelolaan Sumberdaya Air Di Indonesia*, Makalah dipresentasikan pada Lokakarya Kelembagaan Efektif Dalam Pengelolaan Sumberdaya Air Wilayah Sungai Secara Terpadu, Bukittinggi, 5–7 Mei 2002, Kerjasama Pusat Studi Irigasi-Sumberdaya Air, Lahan dan Pembangunan (PSI-PSDLP), Universitas Andalas, Dinas Penegelolaan Sumberdaya Air Provinsi Sumatera Barat, Direktorat Penatagunaan Sumberdaya Air, Ditjen Sumberdaya Air Departemen Kimpraswil, International Water Management Institute (IWMI), Indonesia.

3

Public participation in the development of a management plan for an international river basin: The Okavango case

Anthony R. Turton and Anton Earle

Introduction

The Okavango River basin is an endoreic river that rises in Angola, passes through a narrow piece of Namibia called the Caprivi Strip, and ends in the Okavango delta in Botswana. It is unusual in that it does not drain into the sea, but ends instead in a large inland desert oasis covering an area of 15,844 km^2, where the water is seemingly "lost" to evaporation and the sands of the Kalahari Desert through a wetland system that is a Ramsar site (Ashton and Neal 2003; Turton et al. 2003a: 20). The strategic importance of the river becomes significant when one considers that both the downstream riparians (Namibia and Botswana) have no perennial rivers flowing on their sovereign soil, being mostly located in a semi-desert (see Figure 3.1). The Okavango River and its major tributaries function as a linear oasis in the otherwise relatively arid areas of Botswana and Namibia (Ashton 2003: 167). It is one of the last undeveloped rivers in Africa and great value is attached to this condition by ecological special interest groups, most of which are located outside the basin and, in many cases, outside the respective riparian states. It is therefore an internationalized river basin with many more stakeholders than a so-called "normal" basin would have, making it a good example of the problems related to policy-making in an "internationalized" river basin.

Given the fact that water scarcity is a constraint on the economic growth of four of the most developed states in the Southern African

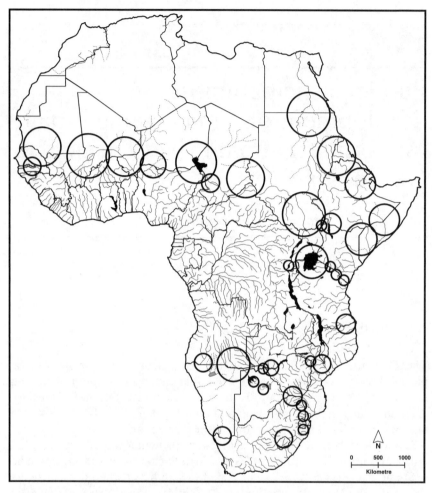

Figure 3.1 Map of perennial rivers in Africa and disputes over water.
Sources: Ashton (2000: 77); Turton et al. (2003a: 10).
Note: The circles indicate known conflicts over water.

Development Community (SADC) region – South Africa, Zimbabwe, Botswana and Namibia – a hydropolitical complex is said to exist because the management of water resources in international river basins forms an issue area of sufficient saliency to influence the patterns of amity and enmity between states (Turton 2003c). Two of the Okavango River basin riparians have been defined as being "pivotal states" in the southern African hydropolitical complex (Namibia and Botswana), and the Okavango River basin has been defined as being an "impacted basin" (Turton 2003a, 2003c; Turton et al. 2003a: 13; 2003b: 28). Water is thus a strategic resource and the management of transboundary systems has the potential for triggering either conflict or cooperation.

This chapter documents some of the key processes that occurred during the life-span of this project, particularly with respect to the evolution of a methodology for the development of a management plan for an international river basin through the process of public participation.

Statement of the problem

The core problem confronting the Permanent Okavango River Basin Water Commission (OKACOM) is encapsulated in four key aspects:
- The two downstream riparians (Namibia and Botswana) are amongst the four most economically developed states in the SADC region, and have water scarcity constraints on their future economic growth potential (Turton 2003a). This raises water resources management to the level of strategic interest and potential "high politics".
- There is no consensus between the three riparian states on a common developmental vision and strategy, making sovereignty one of the fundamental stumbling blocks to potential cooperation (see Turton 2002).
- With a highly variable and relatively small stream-flow (maximum: $16{,}145 \times 10^6$ m^3; minimum: $5{,}321 \times 10^6$ m^3; mean: $9{,}863 \times 10^6$ m^3) (Ashton and Neal 2003: 37), the river simply does not carry a large enough volume of water to satisfy all of the needs of the respective riparian states.
- The hydropolitical relations in the basin are characterized by asymmetry, particularly with respect to institutional development and management capacity among the three riparian states. Angola has been ravaged by civil war, which in turn has severely diminished the administrative capability of the state (Porto and Clover 2003), whereas Namibia and Botswana both have relatively sophisticated administrative capacitates.

Combined, these four core aspects open up the critical need to change the water resources management paradigm away from water-sharing to benefit-sharing instead, if conflict is to be averted in the future.

In search of an appropriate methodology for public participation

Given the relative uniqueness of the social, historical and hydrological context of the Okavango River basin, it was deemed necessary to experiment with a new methodology for public participation in the development of a management plan for the entire basin. The African Water Issues Research Unit (AWIRU) teamed up with Green Cross

International (GCI) through the Water for Peace programme. This in turn was linked with the UNESCO PCCP (From Potential Conflict to Co-operation Potential) initiative. The broad objective was to develop a methodology that was acceptable to the commissioners of OKACOM, and thereby to create experimental space in which they felt comfortable enough to engage with civil society and members of the epistemic community (see Haas 1989, 1992; Haas et al. 1995). Central to this was the recognition that sovereignty is a key issue for states in the developing world, particularly in regions where the colonial experience had been characterized by a liberation struggle (Turton 2002). This meant that any methodology for public participation needed to be based on certain fundamental principles. In this regard the following core principles were used to guide the development of the methodology:

- In the case of international river basins it is governments, and not non-governmental organizations (NGOs), that make decisions. Governments are therefore key stakeholders in the process of decision-making because it is they, and only they, that are accountable to their respective electorates. This recognizes the fundamental principle of sovereignty in international relations involving transboundary rivers (Turton 2002).
- Governments are neither inherently bad nor inherently good. The presumption is thus made that governments want to make the best possible decisions in the circumstances.
- In the context of the developing world, government capacity is generally low and decisions are invariably made against the background of imperfect knowledge (Turton 2003b: 88; Turton et al. 2003b: 71). This has the potential rapidly to escalate tensions in an international river basin when perceptions of threat are couched in terms of strategic interests, such as those arising from water-scarcity limitations to the future economic growth potential of the state.
- This means that decision-making capacity will be improved if government officials are engaged in a neutral manner by exposing them to civil society interests and technical knowledge from the epistemic community.

To meet these objectives, a series of interventions were planned and executed. Each intervention had a clearly defined objective and output. The best way to understand the process is to view it as a series of events, as detailed below.

Event 1: Southern Okavango Integrated Water Development Project

The first relevant event occurred in the 1980s when the government of Botswana decided to launch what was known as the Southern Okavango Integrated Water Development Project (SOIWDP). The core idea behind this project was to reduce evaporative losses in the delta by limiting the area of floodplain in an attempt to develop the water resource base as a viable supply to the mining operation at Orapa (Scudder et al. 1993; Heyns 2003). Central to the project was the dredging of the Boro distributary in order to make it deeper and wider, and thereby to reduce the flooding and consequent evaporative losses. This project was vigorously opposed by a number of special interest groups and resulted in the early internationalization of the Okavango River basin. Central to the opposition to the SOIWDP was the mobilization of scientific knowledge that was used to oppose the government of Botswana. That scientific knowledge was captured in a book *The IUCN Review of the Southern Okavango Integrated Water Development Project*, which was widely distributed by IUCN – The World Conservation Union (see Scudder et al. 1993). After a vigorous set of interactions between the two main players at the time – the government of Botswana and a cluster of special interest groups under the broad umbrella of the IUCN – the SOIWDP was aborted in 1992 (Scudder et al. 1993: xxxi; Heyns 2003: 17), further preventing environmental damage caused by dredging (Scudder et al. 1993: 13).

After this event, little happened in the Okavango basin in terms of development of the water resources, but the foundation of the future hydropolitical dynamics had been laid. Those dynamics were firmly grounded in outside special interest groups, capable of mobilizing significant scientific knowledge and political pressure, with the stated objective of opposing development plans that they felt to be environmentally damaging. In other words, these special interest groups became the custodians of the aquatic ecosystem, but with special emphasis being placed on the Okavango delta only. Another important consequence of this set of hydropolitical dynamics was the emergence of a degree of suspicion on the side of government of the motives and strategies of special interest groups, because the initial hydropolitical configuration had been adversarial in structure. This is particularly relevant in the context of states that had only recently been given independence and that tend to guard their sovereignty jealously.

Event 2: Namibian plans to develop a pipeline

Shortly after Namibia attained its independence in 1990, the government established a number of river basin institutions with co-riparian states (Pinheiro et al. 2003: 114; Turton 2004). These included the Permanent Okavango River Basin Water Commission in 1994 (Treaty 1994).

At the first meeting of OKACOM (Pinheiro et al. 2003: 115; Heyns 1999), the government of Namibia formally announced its intention to develop a pipeline from the Okavango River, starting from an abstrac-

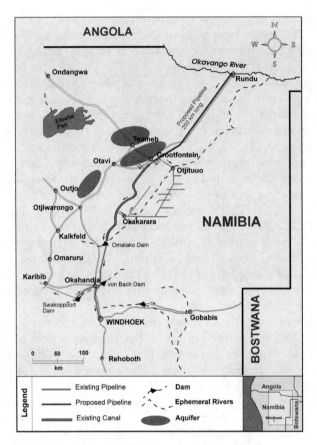

Figure 3.2 The Eastern National Water Carrier in Namibia.
Source: Pinheiro et al. (2003: 113).

tion point near Rundu, feeding water down into the Eastern National Water Carrier (ENWC), and ultimately supplying the capital city Windhoek (see Figure 3.2). This pipeline had been planned as part of the strategic water supply system of the country when it was still being administered by South Africa under a United Nations mandate, so the idea of the pipeline was not new. In fact, the construction of the ENWC began in 1969, initially deriving its water from the Cunene River, but with the stated intention of eventually linking into the Okavango River (Turton 2004: 282; Davies et al. 1993: 167; Davies and Day 1998: 296–299; Heyns 1995: 10). The need for the pipeline was acute however, because Namibian independence coincided with a significant drought, and water resource scarcity was identified as a limiting factor on the future economic growth potential of the state (Heyns 2003: 18).

Reaction to these plans within OKACOM was vigorous, particularly from the downstream riparian Botswana (*Weekly Mail & Guardian* 1996a, 1996b; *Electronic Mail & Guardian* 1997; Ramberg 1997). In an attempt to protect its interests, and presumably having learned from the SOIWDP experience, the government of Botswana registered the delta as a Ramsar site (Jansen and Madzwemuse 2003: 143). Although no official pronouncements have been made regarding the strategic planning behind this registration, it appears that Botswana was trying to use the status as a protected wetland to strengthen its case against Namibian intentions to develop the pipeline. Botswana had seemingly learned the lessons from the SOIWDP experience, and now wanted to use the force of special interest groups to oppose Namibian plans with as much vigour as they had previously opposed Botswana's plans.

One of the results of this set of hydropolitical dynamics was the negative reaction from special interest groups to Namibian plans, specifically regarding the perceived impacts that the pipeline would have on the Okavango delta (Heyns 2003: 18). This negative reaction is growing stronger with the recently announced plans by Namibia's national power utility (NamPower) to develop a small hydropower plant at Popa Rapids in the Caprivi Strip. This is now providing two distinct focal points around which international special interest groups can focus their energies, to the probable detriment of Namibian national interest. The government of Botswana can therefore ease off in its open opposition to the Namibian proposals and leave the special interest groups to do its work for it.

Event 3: Namibian reaction to Botswana's strategy

Being confronted by a debilitating drought, Namibia reacted to Botswana's perceived strategy by launching two specific initiatives. The first was the commissioning of a detailed environmental impact assessment (EIA). This was conducted by the Council for Scientific and Industrial Research (CSIR), an internationally recognized institution with a high level of credibility and integrity. This study found that, although there would be an environmental impact, currently available scientific tools were incapable of measuring the area by which the delta would be reduced (CSIR 1997a, 1997b). In addition to this, the impact could be significantly reduced if the water abstraction took place on the receding portion of the hydrograph. Significantly, however, the study found that two crucial components of the ecological functioning of the delta were flooding (known technically as the Flood Pulse Concept – Davies et al. 1993: 10, 94; Junk et al. 1989; Puckridge et al. 1993; Turton 1999; McCarthy et al. 2000) and sediment transportation.

The second initiative was the registration of a plan with the SADC Water Sector Coordinating Unit (WSCU) that is designed to determine the feasibility of transferring water from the Congo River basin into the Okavango and Zambezi River basins (Heyns 2002: 164). The core argument that underlies this proposed development is that, if Botswana objects to the reduced volume caused by abstracting water via the proposed pipeline, then that volume will be augmented from the Congo River and used by Namibia as strategic needs dictate. In other words, if Namibia puts a given volume of water into the Okavango River from another basin and then abstracts that same volume further downstream, then the nett flows into the delta will remain unchanged (at least in theory, but certainly not in practice given the ecological ramifications related to this practice – in this regard Namibia is opening itself to a third focal point for the mobilization of special interest groups). This logic has not been officially stated in any document, but it is central to any understanding of the hydropolitical dynamics of the Okavango River basin.

At this stage of the hydropolitical history of the Okavango River basin, the dynamic interaction of the two downstream riparian states was based on the core issue of sharing water, and the prevailing trend was clearly towards conflict because there is relatively little water to be shared in the first place, and any upstream abstractions would have a negative impact on the delta downstream. This situation will be exacerbated when Angola starts to abstract large volumes of water for post-war reconstruction. Because this will have a severe impact on both downstream riparians, this fact alone acts as a potential catalyst for cooperation and may end the prevailing adversarial relationship between Namibia and Botswana.

Event 4: Green Cross International Water for Peace intervention

With the hydropolitical dynamics in the Okavango River basin clearly on a trajectory towards conflict, but with the possibility of cooperation arising in the form of the narrow window of opportunity that has been created by the outbreak of peace in Angola, Green Cross International (GCI) decided to focus a component of its Water for Peace programme on the basin. After contracting the African Water Issues Research Unit (AWIRU) at the University of Pretoria to manage the project, some detailed planning was done. This planning was based on the core concepts noted at the start of this chapter, the most important being the clear recognition that it is government and only government that makes binding decisions in international river basins.

The GCI/AWIRU initiative launched a series of workshops in the Okavango River basin with the objectives of (a) isolating the key drivers of the hydropolitical processes in order to make them understandable to

Figure 3.3 Participants at the Green Cross International Water for Peace Workshop in Maun, Botswana.
Note: The OKACOM commissioners are sitting with Sir Ketumile Masire, the former Botswana President, in the centre.

all interested and affected parties; (b) engaging OKACOM commissioners in this process; and (c) creating experimental space in which the prevailing paradigm of water-sharing could be interrogated to the extent that it could be shifted to a new paradigm of benefit-sharing instead.

The first workshop was held in Maun, Botswana, and was attended by OKACOM commissioners from all three riparian states. The scene was set by initially taking all participants out onto the Okavango delta in local *makhoros* (dugout boats), which provided all participants in the workshop with some insights into the complexity of the ecosystem within the Okavango delta. This proved to be a valuable element in the process, because the Angolan commissioner had never been to the delta before and thus had no real knowledge of the significance of the aquatic ecosystem as a provider of ecosystem services other than merely a water resource.

In addition to this, the best available scientists were invited to present papers on carefully selected topics. A core component of the strategy was to invite the three riparian states to present position papers in order to lay the foundations for understanding the needs and expectations of the three riparian states. The OKACOM commissioners declined the offer to present individual papers, and chose instead to present a joint paper. This was seen as an encouraging sign by GCI and AWIRU.

The facilitator of the process at that time (Anthony Turton) had arranged with the BBC to send a TV cameraman to the workshop. An agreement had been reached with the BBC cameraman that he would not harass any of the commissioners and he would not try to trick them into making statements. This agreement was presented to the commissioners, who were informed that they were under no obligation to speak with the BBC cameraman but, if they wished to do so, then the opportunity existed for them to say whatever they wished to communicate to the world at large. The cameraman also spent his time shooting a documentary, into which OKACOM statements could be inserted as appropriate.

There were four specific outputs of this workshop:

- OKACOM presented a joint paper on its vision for the management of the entire basin in the future. This was significant because it represented a shift in focus away from the Okavango delta region to the whole river basin. It was also the first meeting of OKACOM commissioners outside of their regular rotation of official engagements.
- All three riparian states used the opportunity provided by the presence of the BBC cameraman. Each made a statement, independent of the others and in most cases with no knowledge of what the others had said. Each of these statements was overwhelmingly positive in its orientation, with a strong commitment to using water for peace. Significantly, the Botswana commissioner recognized Angola's right to use the water in post-conflict reconstruction projects. In similar fashion, the Namibian commissioner acknowledged Botswana's concerns about the impact of Namibia's proposed pipeline, and stated categorically that Namibia was committed to the peaceful resolution of the problem. The Angolan commissioner stated that his government recognized downstream concerns and that they wanted to use water as a catalyst for peace, because for too long they had been living with the bitterness of war. These statements were broadcast by the BBC World Service as part of its coverage of the build-up to the Third World Water Forum. The message thus reached an estimated audience of around 500 million people.
- A set of high-quality scientific papers was generated. These were brought together into the *Proceedings* and made available to all participants. In essence these papers represented a summary of the best available scientific knowledge of the complexities associated with the Okavango River basin.
- The First Generation Strategic Report on the Okavango River Basin was developed in the form of a paper by Peter Ashton and Marian Neal, which summarized the strategic problems in the form of a diagram that was easily understandable to all interested and affected parties (see Figure 3.4).

Armed with the First Generation Strategic Report, and encouraged by the support that the OKACOM commissioners had given to the Green Cross International Water for Peace project, AWIRU took all of the material available and developed what was officially called *An Assessment of*

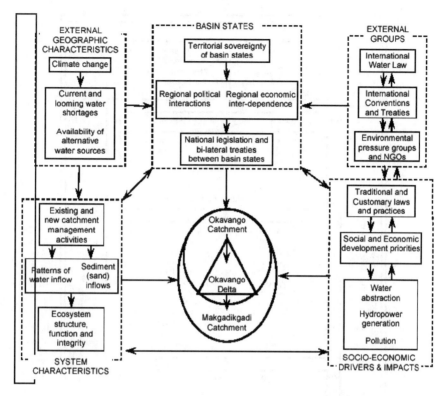

Figure 3.4 Graphical representation of the First Generation Strategic Report on the Okavango River Basin.
Source: Ashton and Neal (2003: 58).

the Hydropolitical Dynamics of the Okavango River Basin. For the purposes of this chapter, it can be called the Second Generation Strategic Report on the Okavango River basin (see Figure 3.5). This synthesized all available knowledge on the Okavango River basin and became an input into the second workshop, which was held at the Gobabeb site of the Desert Research Foundation of Namibia (DRFN). Present at that meeting were seconded representatives of OKACOM from all three riparian states. There was also a strong NGO presence. Included at this time were scientists from the Water Ecosystems Resources in Regional Development (WERRD) project.[1] The intention of GCI and AWIRU was to streamline the Second Generation Report and make it less technical and more user-friendly. In order to achieve this objective, the professional services of Dr Barbara Heinzen were engaged. She is a highly respected facilitator with skills in the field of strategic scenario planning, and her brief was to use the material provided to start developing a set of scenarios that all the interested and affected parties could relate to.

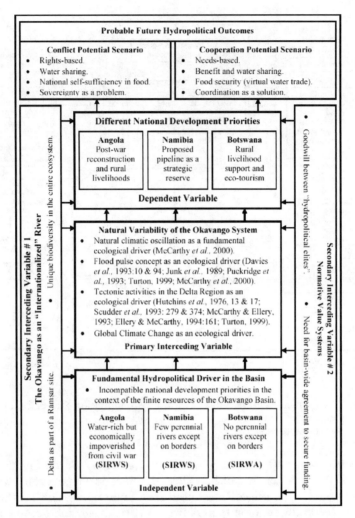

Figure 3.5 Graphical representation of the Second Generation Strategic Report on the Okavango River Basin.
Source: Turton et al. (2003c: 361).

Dr Heinzen broke the participants into four groups, with an OKACOM commissioner in each group. Each group also contained scientists and NGO representatives. Over a period of two days the task of these four groups was to develop a consensus document that could be encapsulated in one graphic image, using all available scientific knowledge but specifically based on the Second Generation Strategic Report on the Okavango River Basin. The output of this process can be called the Third Genera-

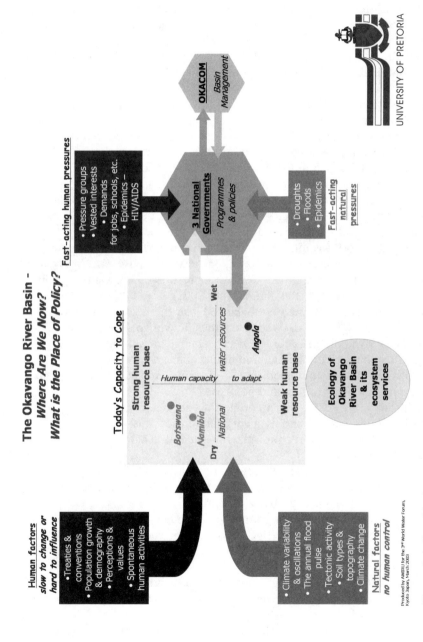

Figure 3.6 Graphical representation of the Third Generation Strategic Report on the Okavango River Basin, presented at the Third World Water Forum in Japan. *Source*: Green Cross International.

tion Strategic Report on the Okavango River Basin and is presented in Figure 3.6. This was taken to the Third World Water Forum in Kyoto in 2003 and presented in an appropriate forum there. The honour associated with having the Okavango River basin case presented at such a prestigious global event acted as somewhat of a stimulant to the participants, because they felt that they were being given a voice.

What the Third Generation Strategic Report on the Okavango River Basin contained in one powerful graphic image were the following core issues:
- There were human forces and natural forces at work at various points within the overall river basin.
- These two categories of force could be divided into slow-acting forces over which no human control was possible (shown on the left of the graphic), and fast-acting forces over which some degree of human control was imaginable (shown on the right of the graphic).
- These forces were acting upon the different riparian states in different ways, having been mediated through what was described as "today's capacity to cope". This second-order resource focus (the adaptive capacity of society) was different for each riparian, with Angola having a lot of water resources but a weak human resource base, and Botswana and Namibia having relatively few water resources but a stronger human resource base (Turton and Warner 2002).
- Today's capacity to cope was "balanced" on the ecological goods and services that could be derived from the Okavango River basin. This balance is dynamic in nature and can change rapidly in a non-linear way in response to the human and natural factors at work.
- All of these issues combined have a dynamic impact on the national government's programmes and policies.
- Significantly, OKACOM, as an organ that has been created by all three riparian states, can affect only a very limited set of issues.

The GCI/AWIRU intervention thus succeeded in achieving the following specific objectives:
- It generated a high degree of credibility with OKACOM.
- It showed OKACOM that not all interactions with special interest groups need necessarily be bad or painful.
- It showed that the epistemic community is indeed capable of providing the level of scientific knowledge needed, in a readily digestible format and in a way that can cause a fundamental rethinking of the core problem being managed.
- It showed that water-sharing is probably not sustainable and is likely to lead to conflict, whereas benefit-sharing is viable and will probably lead to cooperation instead.
- In order for benefits to be shared in a way that is fair and equitable, solutions need to be sourced at a high level of strategic thinking and

planning. This implies that strategic environmental assessment (SEA) skills are an essential element of future management at the basin level.

Event 5: Universities Partnership for Transboundary Waters

Taking the outputs of the Green Cross International Water for Peace project as a foundation of cumulative scientific knowledge in water resources management, the Universities Partnership for Transboundary Waters (UPTW)[2] hosted a workshop at Oregon State University from 14 to 16 April 2003. Under the official title of "Dialogue on Sub-national Stakeholder Participation in International River Basin Environmental Initiatives – Models, Successes and Failures", the initiative brought together managers from the Okavango, the Lempa and the San Juan river basins, with AWIRU as a co-facilitator. This event was funded by the Carnegie Corporation through the Pacific Institute. It enabled the experience gained in the GCI/AWIRU process to be presented to other river basin commissioners from Central America. Emerging from the resultant dialogue were three clear issues relevant to the Okavango: (a) OKACOM is being hampered because of the absence of a permanent secretariat; (b) interaction with donors is problematic for a variety of reasons; (c) it is difficult to coordinate the needs of the riparian states with the needs and interests of the donor agencies. Significantly, the UPTW/AWIRU initiative succeeded in delivering a firm result to OKACOM, which felt that its problems were starting to be aired in a way that could realistically lead to their resolution.

Event 6: National Heritage Institute Sharing Waters project

In similar vein to the UPTW case, the National Heritage Institute Sharing Waters: Towards a Transboundary Consensus on the Management of the Okavango River project is currently ongoing. Involving a consortium of partners including the IUCN Regional Office for Southern Africa (ROSA), and funded by the US Agency for International Development (USAID), this will accomplish a range of objectives. One of these is to take the goodwill generated by earlier work and translate that into capacity-building within the basin in a sustainable way.

Event 7: Woodrow Wilson Center Project

The Woodrow Wilson International Center for Scholars in Washington DC is running a project called Environmental Change and Security.[3] One element of that project is the Water Working Group. AWIRU has facilitated cooperation with OKACOM and has brought together the Woodrow Wilson Center Water Working Group and GCI in what is officially entitled the "Okavango Focus Meeting". This was held in the delta from 23 to 25 October 2003. It was well received and made a valuable contribution to consolidating the position of the Third Generation Strategic Report on the Okavango River Basin, taking it to a new level of sophistication. Significantly, there was strong evidence at that meeting that the Third Generation Strategic Report has acted as a catalyst to the local stakeholders, who showed evidence of strong buy-in to the contents of the document. This exceeded the expectations of the organizers.

Event 8: Water Ecosystems Resources in Regional Development project

Since the Gobabeb workshop in which the Second Generation Strategic Report was developed into the Third Generation document, WERRD has shown an interest in the GCI/AWIRU initiative. One of the tangible elements of this has been a slight shift in focus for the WERRD project to include scenarios about future resource use in the Okavango River basin. The WERRD project eventually adopted some of the elements contained in the Third Generation Strategic Report, to the extent that the notion of future scenarios for the basin was contained in the final report.[4]

Conclusion

The Green Cross International Water for Peace project has shown that public participation can become a meaningful part of river basin management in the international sphere. More importantly, the GCI/AWIRU intervention has shown that trust is a vital element in the relationship between government and the epistemic community. The hydropolitical history of the Okavango River basin has a period in which a highly adversarial relationship existed between special interest groups and government. This is ongoing and has had a strong impact that has tended to make government suspicious of the motives of special interest groups.

A major achievement of the GCI/AWIRU initiative has been the change in paradigm away from sharing water to sharing benefits instead. This is deeply encouraging and is relevant to a number of international river basins that are characterized by a high level of contestation and a low level of resource availability, such as the Nile, Orange, Limpopo, Incomati and Maputo, to name but a few. In fact, it is relevant to all of the river basins that Wolf et al. (2003) have defined as being "at risk". The change in focus, away from broad aspirations to a more clearly defined set of realistic feasibilities instead, is a characteristic of sustainable river basin management, in the sense that it represents a shift away from what should be done to what can be done. The project also represents the expansion of the management focus away from simply a delta management plan, to a plan that encompasses the entire river basin and beyond, as the sharing of benefits is sourced from outside.

The support of third-party actors as honest brokers is also highly relevant. The key impact of the GCI/AWIRU initiative can be measured in the significant changes between the First, Second and Third Generation Strategic Reports on the Okavango River Basin. Each evolution has become increasingly nuanced and has been accompanied by a higher level of legitimacy for the core elements than was evident in the efforts of the scientific community alone. This shows that legitimacy is a quality that is given to the basin management plan by key players such as government, via a process of engagement in which the epistemic community is involved in a non-threatening manner.

Acknowledgements

We would like to acknowledge the Tokyo University of Agriculture and Technology (TUAT), OKACOM, Green Cross International, the Council for Scientific and Industrial Research (CSIR), the Woodrow Wilson International Center for Scholars, Arcus-Gibb (Pty) Ltd, the Water Ecosystems Resources in Regional Development project (WERRD), the National Heritage Institute (NHI), the World Conservation Union (IUCN), the Carnegie Corporation, the Pacific Institute, USAID, the Desert Research Foundation of Namibia (DRFN) and the Universities Partnership for Transboundary Waters (UPTW) for their support for the work that is ongoing and that is reflected in this chapter. We alone accept responsibility for any assessment or pronouncements made.

Notes

1. See ⟨http://www.okavangochallenge.com/okaweb/⟩.
2. See ⟨http://waterpartners.geo.orst.edu/⟩.
3. See ⟨http://ecsp.si.edu⟩.
4. The reader is urged to look at ⟨http://www.okavangochallenge.com/okaweb/⟩ for further details as the project evolves.

REFERENCES

Ashton, P. (2000) "Southern African Water Conflicts: Are They Inevitable or Preventable?", in H. Solomon and A. R. Turton (eds), *Water Wars: Enduring Myth or Impending Reality?*, African Dialogue Monograph Series No. 2. Durban: ACCORD Publishers.

Ashton, P. (2003) "The Search for an Equitable Basis for Water Sharing in the Okavango River Basin", in M. Nakayama (ed.), *International Waters in Southern Africa*. Tokyo: United Nations University Press.

Ashton, P. and M. Neal (2003) "An Overview of Key Strategic Issues in the Okavango Basin", in A. R. Turton, P. Ashton and T. E. Cloete (eds), *Transboundary Rivers, Sovereignty and Development: Hydropolitical Drivers in the Okavango River Basin*. Pretoria and Geneva: AWIRU and Green Cross International.

CSIR [Council for Scientific and Industrial Research] (1997a) *The Okavango River – Grootfontein Pipeline Link to the Eastern National Water Carrier (E.N.W.C.) in Namibia: An Assessment of the Potential Downstream Environmental Impacts in Namibia and Botswana*. Pretoria: Council for Scientific and Industrial Research.

——— (1997b) "An Assessment of the Potential Downstream Impacts in Namibia and Botswana of the Okavango River – Grootfontein Pipeline Link to the Eastern National Water Carrier in Namibia: Initial Environmental Evaluation Report", Contract Report to Water Transfer Consultants, Windhoek, Namibia by Division of Water, Environment and Forestry Technology, CSIR. Report No. ENV/P/C 97120.

Davies, B. R. and J. Day (1998) *Vanishing Waters*. Cape Town: University of Cape Town Press.

Davies, B. R., J. H. O'Keefe and C. D. Snaddon (1993) *A Synthesis of the Ecological Functioning, Conservation and Management of South African River Ecosystems*, Water Research Commission Report No. TT 62/93.

Electronic Mail & Guardian (1997) "Namibia Relations Get Frostier with Botswana over Okavango", 28 January.

Ellery, W. N. and T. S. McCarthy (1994) "Principles for the Sustainable Utilization of the Okavango Delta Ecosystem, Botswana", *Biological Conservation* 70(2): 159–168.

Haas, P. M. (1989) "Do Regimes Matter? Epistemic Communities and Mediterranean Pollution Control", *International Organization* 43(3): 377–403.

——— (1992) "Introduction: Epistemic Communities and International Policy Coordination", *International Organization* 46(1): 1–35.

Haas, P. M., R. O. Keohane and M. Levy (1995) *Institutions of the Earth: Sources of Effective International Environmental Protection*. Cambridge, MA: MIT Press.

Heyns, P. S. (1995) "Existing and Planned Development Projects on International Rivers within the SADC Region", in *Proceedings of the Conference of SADC Ministers Responsible for Water Resources Management*, Pretoria, 23–24 November.

——— (1999) *Water Resources Management in Namibia with Regard to the Implementation of the Protocol on Shared Watercourse Systems of the Southern African Development Community*. Windhoek: Department of Water Affairs.

——— (2002) "Interbasin Transfer of Water between SADC Countries: A Development Challenge for the Future", in A. R. Turton and R. Henwood (eds), *Hydropolitics in the Developing World: A Southern African Perspective*. Pretoria: African Water Issues Research Unit (AWIRU).

―――― (2003) "Water Resources Management in Southern Africa", in M. Nakayama (ed.), *International Waters in Southern Africa*. Tokyo: United Nations University Press.

Hutchins, D. G., S. M. Hutton and C. R. Jones (1976) "The Geology of the Okavango Delta", in *Proceedings of the Symposium on the Okavango Delta and its Future Utilization*, National Museum, Gaborone, Botswana, Botswana Society.

Jansen, R. and M. Madzwemuse (2003) "The Okavango Delta Management Plan Project: The Need for Environmental Partnerships", in A. R. Turton, P. Ashton and T. E. Cloete (eds), *Transboundary Rivers, Sovereignty and Development: Hydropolitical Drivers in the Okavango River Basin*. Pretoria and Geneva: AWIRU and Green Cross International.

Junk, W. J., P. B. Bayley and R. E. Sparks (1989) "The Flood Pulse Concept in River-floodplain Systems", in D. P. Dodge (ed.), *Proceedings of the International Large Rivers Symposium (LARS)*, Canadian Special Publication of Fisheries and Aquatic Sciences, No. 106, pp. 110–127.

McCarthy, T. S. and W. N. Ellery (1993) "The Okavango Delta", *Geobulletin* 36(2): 5–8.

McCarthy, T. S., G. R. J. Cooper, P. D. Tyson and W. N. Ellery (2000) "Seasonal Flooding in the Okavango Delta, Botswana – Recent History and Future Prospects", *South African Journal of Science* 96: 25–33.

Pinheiro, I., G. Gabaake and P. Heyns (2003) "Cooperation in the Okavango River Basin: The OKACOM Perspective", in A. R. Turton, P. Ashton and T. E. Cloete (eds), *Transboundary Rivers, Sovereignty and Development: Hydropolitical Drivers in the Okavango River Basin*. Pretoria and Geneva: AWIRU and Green Cross International.

Porto, J. G. and J. Clover (2003) "The Peace Dividend in Angola: Strategic Implications for Okavango Basin Cooperation", in A. R. Turton, P. Ashton and T. E. Cloete (eds), *Transboundary Rivers, Sovereignty and Development: Hydropolitical Drivers in the Okavango River Basin*. Pretoria and Geneva: AWIRU and Green Cross International.

Puckridge, J. T., F. Sheldon, A. J. Boulton and K. F. Walker (1993) "The Flood Pulse Concept Applied to Rivers with Variable Flow Regimes", in B. R. Davies, J. H. O'Keefe and C. D. Snaddon (eds), *A Synthesis of the Ecological Functioning, Conservation and Management of South African River Ecosystems*, Water Research Commission Report No. TT 62/93.

Ramberg, L. (1997) "A Pipeline from the Okavango River?", *Ambio* 26(2): 129.

Scudder, T., R. E. Manley, R. W. Coley, R. K. Davis, J. Green, G. W. Howard, S. W. Lawry, P. P. Martz, P. P. Rogers, A. R. D. Taylor, S. D. Turner, G. F. White and E. P. Wright (1993) *The IUCN Review of the Southern Okavango Integrated Water Development Project*. Gland, Switzerland: IUCN Communications Division.

Treaty (1994) "Agreement between the Governments of the Republic of Angola, the Republic of Botswana and the Republic of Namibia on the Establishment of a Permanent Okavango River Basin Water Commission (OKACOM)", Signatory Document, signed by Representatives of the Three Governments, Windhoek, 15 September 1994.

Turton, A. R. (1999) "Sea of Sand, Land of Water: A Synopsis of Some Strategic

Developmental Issues Confronting the Okavango Delta", MEWREW Occasional Paper No. 6. Water Issues Study Group, School of Oriental and African Studies (SOAS), University of London; available at ⟨http://www.soas.ac.uk/Geography/WaterIssues/OccasionalPapers/home.html⟩.

—— (2002) "Water and State Sovereignty: The Hydropolitical Challenge for States in Arid Regions", in A. Wolf (ed.), *Conflict Prevention and Resolution in Water Systems*, pp. 516–533. Cheltenham: Edward Elgar.

—— (2003a) "Environmental Security: A Southern African Perspective on Transboundary Water Resource Management", in *Environmental Change and Security Project Report*, Woodrow Wilson Center, Issue 9. Washington, DC: Woodrow Wilson International Center for Scholars.

—— (2003b) "The Hydropolitical Dynamics of Cooperation in Southern Africa: A Strategic Perspective on Institutional Development in International River Basins", in A. R. Turton, P. Ashton and T. E. Cloete (eds), *Transboundary Rivers, Sovereignty and Development: Hydropolitical Drivers in the Okavango River Basin*. Pretoria and Geneva: AWIRU and Green Cross International.

—— (2003c) "The Political Aspects of Institutional Development in the Water Sector: South Africa and Its International River Basins", unpublished draft of a D.Phil. thesis, Department of Political Science, University of Pretoria, Pretoria.

—— (2004) "Evolution of Water Management Institutions in Select Southern African International River Basins", in A. K. Biswas, O. Unver and C. Tortajada (eds), *Water as a Focus for Regional Development*. London: Oxford University Press.

Turton, A. R. and J. Warner (2002) "Exploring the Population/Water Resources Nexus in the Developing World", in G. D. Dabelko (ed.), *Finding the Source: The Linkage between Population and Water*, Environmental Change and Security Project (ECSP). Washington, DC: Woodrow Wilson Center.

Turton, A. R., P. Ashton and T. E. Cloete (2003a) "An Introduction to the Hydropolitical Drivers of the Okavango River Basin", in A. R. Turton, P. Ashton and T. E. Cloete (eds), *Transboundary Rivers, Sovereignty and Development: Hydropolitical Drivers in the Okavango River Basin*. Pretoria and Geneva: AWIRU and Green Cross International.

Turton, A. R., A. Nicol and J. A. Alan (2003b) *Policy Options for Water Stressed States: Emerging Lessons from the Middle East and Southern Africa*. Pretoria and London: AWIRU and Overseas Development Institute.

Turton, A. R., P. Ashton and T. E. Cloete (2003c) "Hydropolitical Drivers and Policy Challenges in the Okavango River Basin", in A. R. Turton, P. Ashton and T. E. Cloete (eds), *Transboundary Rivers, Sovereignty and Development: Hydropolitical Drivers in the Okavango River Basin*. Pretoria and Geneva: AWIRU and Green Cross International.

Weekly Mail & Guardian (1996a) "Plan Could Turn Okavango to Dust", 29 November.

—— (1996b) "Namibia Almost Certain to Drain Okavango", 6 December.

Wolf, A. T., S. B. Yoffe and M. Giordano (2003) "International Waters: Identifying Basins at Risk", *Water Policy* 5(1): 29–60.

4
Transboundary environmental impact assessment as a tool for promoting public participation in international watercourse management

Jessica Troell, Carl Bruch, Angela Cassar and Scott Schang

Access to an adequate supply of quality freshwater is essential to both human and ecosystem health and well-being. Yet over 1 billion people worldwide lack access to safe drinking water, and 3.4 million people, the majority of them children in the developing world, die annually from water-related diseases (Scanlon et al. 2003). Water scarcity also stands as a major impediment to poverty alleviation, because water is a fundamental input for key economic sectors and sustains the ecological services necessary to support communities that rely on subsistence and resource-based economies. The world community has included among the Millennium Development Goals (MDGs) a commitment to halving, by the year 2015, the number of people without sustainable access to safe drinking water (World Bank Group 2004). Improved planning, regulation and management of freshwater resources will be essential if we are to reach these goals in an environmentally sustainable manner (WHO 2004). With 261 river basins shared by two or more states, one of the most complex aspects of this challenge is the management of shared watercourses.

Over the past three and a half decades, several resource-specific treaties and river basin institutions have emerged throughout the world seeking to facilitate cooperation among riparian states through improved technical and administrative management of shared waters. At the same time, there has been an increasing international awareness of the critical role of the public in protecting the environment and the need to involve the public in environmental decision-making. Recent evolution in international law regarding public participation in the management of shared watercourse systems and the widespread inclusion of relevant provisions in both regional and water-body-specific instruments indicate that norms related to public involvement are beginning to crystallize with respect to the management of transboundary waters (Bruch 2003b).

This chapter focuses on transboundary environmental impact assessment (TEIA), a planning tool that has the potential to implement the norms of participatory management in the context of international water resources. TEIA is a process in which governments, international institutions and the public assess the likely or potential environmental (and often social and economic) impacts of a proposed activity. At both the regional and international levels, treaties and more informal mechanisms are emerging that include TEIA as a means for taking both a precautionary and a participatory approach to planning activities with potential transboundary impacts. If structured and implemented appropriately, these legal mechanisms can provide a practical means for facilitating and enhancing public participation in decision-making related to shared watercourses.

As will be elaborated below, public participation in the TEIA process proffers a number of benefits for both the affected public and the decision makers conducting the assessment. However, these benefits appear to be largely unrealized at present. There are few effective operational frameworks for TEIA, and in many places the infrastructure does not yet exist to encourage meaningful public participation in the management of international watersheds. The growing number of regional TEIA initiatives offers hope for the future implementation of public participation in international impact assessment. Additional directed research that analyses the implementation of public participation practices in the various TEIA mechanisms over time could help to ensure that the full benefits from those practices are realized. This chapter provides an overview of emerging practice and the potential for realizing these benefits.

Background

The public's role in environmental impact assessment

Environmental impact assessment (EIA) is "[a]n assessment of the likely or potential environmental impacts of [a] proposed activity" (UNEP 1987). Rather than responding to environmental impacts as they occur, the EIA process enables decision makers to anticipate the consequences of their actions and avoid or minimize adverse effects. Impact assessment is aimed not necessarily at requiring specific environmental outcomes, but rather at ensuring a more open and inclusive decision-making process to arrive at a better substantive result. A significant element of this precautionary approach is public disclosure of information regarding activities that have potential environmental impacts and the solicitation of public input at various stages of the EIA.

Since its introduction in the United States' National Environmental Policy Act of 1969 (NEPA), the concept of EIA has spread rapidly, with countries around the world adopting EIA laws, procedures and institutions (OECC 2000). Although differing political regimes, regional environmental priorities and cultural values have contributed to variations in EIA processes and standards, the general elements of the EIA process are relatively consistent, at least in principle (Preiss 1999; Timoshenko 1988).

Public involvement in the assessment process is one element common to almost all EIA regimes. The level of public participation in EIA and, in turn, the meaningfulness of that participation to both the assessors and the relevant public vary greatly, however. Some impact assessment regimes tend to postpone public involvement and consult designated "relevant" stakeholders only when much of the substantive decision-making has already been accomplished. Other systems involve the public earlier on in the assessment process, and enable the public to influence decision-making in a more substantive manner. Thus, it is important to note whether the public are involved in the initial "screening" step of an EIA, in which a preliminary assessment is undertaken to determine whether the proposed project triggers the EIA requirements, or are to be provided an opportunity to review and comment at the "scoping" stage of an impact assessment, in which the party preparing the EIA determines which impacts, alternatives and mitigation measures should be assessed in the EIA.[1] Perhaps the most widespread mechanism for involving the public in the EIA process in practice, however, is to make the draft EIA available to the public for comment. The means for eliciting comments range from publication in a government gazette to active dissemination in local communities without access to media, accompanied by a public meeting or even individual interviews. EIA systems also differ in the timing and length of the public comment period and whether and how comments should be accounted for in the final EIA.

At its core, an EIA is about gathering information and exploring alternatives to ensure that the impacts of proposed developments on the environment are understood, acceptable and managed appropriately. Local communities and non-governmental organizations (NGOs) often have detailed knowledge of their local environment that is not available to governments or institutions making the policy decisions that affect those areas, including traditional knowledge that conventional approaches often overlook.[2] Thus, involving the public in the assessment process is a vital means for widening the potential sources of relevant information, such as supplementary baseline data about local environmental conditions and processes and clarification of the values and trade-offs associated with the various alternatives from the affected populations (UNEP 2002).

The failure to involve the public appropriately in EIA, on the other hand, can contribute to public resistance to the project, increased administrative costs and a poorly designed and executed project. For example, the construction of the Pak Mun Dam on a tributary to the Mekong River in Thailand did not include public participation in the assessment process, resulting in objections by affected communities that the compensation they were offered was inadequate. The unexpected costs involved in addressing these objections increased the dam's overheads, altering the cost–benefit analysis for the final assessment decision (Bruch 2003b).

Public participation and transboundary environmental impact assessment

Simply stated, a TEIA is an EIA that is performed when environmental impacts have the potential to affect a state other than the one in which the environmental harm or the project that results in the harm originates. Although the general structure of a TEIA has many elements in common with a domestic EIA (including public participation and the overall chronology of stages described above), a TEIA imposes additional political, administrative and regulatory layers on the process, making it more complex than the EIA process (Knox 2002).

As discussed with regard to EIAs above, public participation in TEIAs brings both benefits and costs. There are several additional constraints often faced by proponents, responsible authorities and stakeholders at the national level that are exacerbated when transboundary participation is involved. These include:
- resource constraints and the extra costs of involving additional, and sometimes very different, sets of stakeholders, especially from other countries;
- legislative and regulatory differences in national EIA processes;
- access to justice (which, in the transboundary context, can raise constitutional issues such as standing, distance and scale);
- institutional and cultural differences in decision-making and participation;
- language barriers and a need for translation;
- varying levels of education and literacy across borders;
- varying sensitivity to traditionally marginalized populations;
- physical remoteness of stakeholders; and

- the significant time it may take to mount a meaningful TEIA public participation process across borders, languages and cultures.

Yet, with the added complications presented by conducting an impact assessment on transboundary waters, there are also additional benefits to be reaped by conducting a TEIA appropriately and with the participation of the public affected on both sides of the border. In the transboundary context, stakeholders can include the national, state/provincial and local governments of the riparian countries involved in the TEIA. As such, a properly conducted TEIA can open lines of communication between nations and strengthen relationships between governments, engendering more cooperative and proactive management of transboundary water resources and potentially averting conflict. When international institutions (such as river basin organizations) undertake TEIA, they gain valuable experience at the project level with participatory management that can then be translated to all levels of water governance in which the institution is involved. Further, when the citizens on both sides have had a chance to be heard, the resulting decisions may reflect local knowledge and experience to the benefit of all.

Over recent decades, TEIA procedures have emerged through a patchwork of treaties, declarations and customary law.[3] Owing to the cross-border nature of the impacts it seeks to address, TEIA has emerged most concretely through regional initiatives. Thus, to a significant extent, the current status of TEIA may be best understood by comparing the various regional articulations of TEIA principles and approaches to public participation. In the next section we provide an overview of a variety of regional TEIA approaches.

Regional TEIA initiatives

TEIA in Asia

Although Asia currently lacks a formalized TEIA system, the basin countries of the Mekong River are making significant progress towards TEIA in the management of that watercourse.

The Mekong River Commission (MRC) has started to consider ways to promote both domestic EIA within individual states and TEIA in basin-wide river management. Draft technical guidelines and policy advice for a TEIA system were completed in May 2002, and the MRC is currently working with the national Mekong committees to develop the guidelines further and suggest potential procedures and protocols appropriate to each of the member states (MRC 2002a).

The riparian nations also have some historical experience in dealing with EIA and TEIA. For example, the 1995 Agreement on the Cooperation for the Sustainable Development of the Mekong River Basin between Cambodia, Lao PDR, Thailand and Viet Nam requires the riparian nations to provide timely notification and consultation prior to implementing any projects using the river (MRC 1995). Although not directly referring to TEIA or EIA, the substantive requirements are similar, including the obligation to evaluate and discuss the potential impacts of a proposed use of the river.

In addition to the MRC's efforts to promote TEIA and EIA along the river, there are efforts to promote public participation in the region more broadly. These include a set of guidelines for the application of public participation principles in the context of the MRC, a proposed regional framework for ensuring transparency, public participation and accountability, and ongoing efforts by an NGO coalition (MRC n.d.; Nicro et al. 2002; TAI n.d.).[4]

TEIA in Africa

East African lawmakers are in the process of enacting TEIA provisions that may be a useful model for other regions where economic and political integration among states is not as extensive as in Europe.

The East African Community

The three member countries of the East African Community (EAC) – Kenya, Uganda and Tanzania – concluded a Memorandum of Understanding (MOU) on Environment Management in late 1998 (EAC 1998). The MOU provides for public involvement in environmental decision-making and harmonization of environmental laws among the EAC states. Specifically, the MOU states that all partner countries are to adopt domestic EIA laws that enable public participation "at all stages of the process". Additionally, the MOU endorses TEIA in international water management through explicit promotion of EIA and harmonization of EIA laws in conjunction with managing shared water resources (especially relating to Lake Victoria), and incorporates non-discrimination provisions that require affected states' citizens to receive no less opportunity to participate in the EIA process than the sponsoring country's citizens. This use of the "non-discrimination" principle, requiring that states of origin provide equivalent opportunities for public participation to the public of affected states as are offered to their own public, appears to be developing as a customary principle of TEIA and is included in both North American and European instruments as well (see below).

The African Centre for Technology Studies, the Environmental Law Institute and other organizations are working with the EAC Secretariat in seeking to implement the various TEIA provisions in the Treaty for the Establishment of the East African Community (EAC 1999) and the environmental MOU signed by the three countries by finalizing the development of EIA guidelines for the shared ecosystems of East Africa (Sikoyo 2005). The guidelines have a detailed annex that seeks to promote public participation in the TEIA process. The guidelines have been approved by the EAC and are expected to be incorporated into a new environmental protocol to the EAC Treaty.

The Southern African Development Community

Following the establishment of the Southern African Development Community (SADC) in 1992, the SADC developed and then updated a Protocol on Shared Watercourses (SADC 1992, 2001). The Protocol requires notification (including results of any EIA conducted) between states when planned measures affecting shared watercourses may have a significant adverse effect on other states (SADC 2001). Thus, although SADC has TEIA requirements for watercourses, the public participation provisions currently are limited to state-level notification.

The African Union

The African Union Assembly adopted a revised African Convention on the Conservation of Nature and Natural Resources in 2003 (Algiers Convention), updating the 35-year-old Convention to include more detailed provisions on public participation and water management. The Convention obliges parties to ensure that EIAs are conducted at the earliest possible stage and includes a number of new provisions seeking to promote broader access to information and public participation.[5] Notably, the Convention incorporates the "non-discrimination" principle discussed above in relation to the EAC Treaty by requiring parties from which a transboundary environmental harm originates to ensure that any person in another party affected by such harm has a right of access to administrative and judicial procedures equal to that afforded to nationals or residents of the party of origin in cases of domestic environmental harm.

TEIA in North America

The North American Agreement on Environmental Cooperation

The primary international framework addressing TEIA in North America is the North American Agreement on Environmental Cooperation (NAAEC), which established the North American Commission on Environmental Cooperation (CEC) (NAAEC 1994). Pursuant to a mandate of the NAAEC, the parties resolved in 1997 to develop a binding Transboundary Environmental Impact Assessment (TEIA Agreement) (CEC 1997).

Despite the goal of completing the instrument by 1998, unresolved questions about the extent of the Agreement's application have stalled negotiations, and the TEIAA remains in draft form (Moreno et al. 1999; Knox 2002). It is therefore difficult to comment with any certainty on the exact extent and scope of the Agreement.

The draft provisions on public participation and notification are well developed, however, and do not appear to be controversial. They include provisions for public notification and solicitation of comments from stakeholders in the potentially affected country in a manner equivalent to opportunities provided to the public in the country of origin, and at an early enough juncture to enable a meaningful public participation process. Additionally, the draft TEIAA would require a party of origin to allow the public of the potentially affected party to submit comments for the TEIA process and participate in any public hearing or meeting related to the TEIA to the same extent accorded to the public of the party of origin.

United States/Canada: The International Joint Commission

The International Joint Commission (IJC) is a bilateral institution established by the United States and Canada under the Treaty Relating to Boundary Waters of 1909. Close inspection reveals that the IJC has pursued TEIA, albeit through an informal process and without explicitly terming it as such.

Article VIII of the Boundary Waters Treaty requires the Parties or members of the affected public to submit "applications" to the IJC for permission of intended "uses, obstructions, and diversions ... affecting the natural level or flow of boundary waters on the other side" of the United States–Canada border. The process of application closely resembles a TEIA procedure. The proponent submits an application for approval first to their relevant government authority, which assesses the need for IJC approval. If deemed necessary, the application is submitted to the IJC, which reviews the application, publishes notice and conducts public hearings before recommending approval or denial of an application.

Operated in good faith, this informal system has been mutually beneficial for both countries, and has provided an opportunity for the public on both sides of the border to be actively involved in decisions that affect their water resources (Paisley 2002).

United States/Mexico: The Border Environment Cooperation Commission and the La Paz Agreement

A side agreement to the North American Free Trade Agreement created the Border Environment Cooperation Commission (BECC) to help oversee environmental infrastructure projects in the border region (BECC

1993). Under the BECC's charter, an environmental assessment must be completed in order to certify infrastructure projects in the United States–Mexico border region. If a project requires an EIA according to the domestic law of the place where the project will be located or executed, the EIA that was submitted to the appropriate domestic authority also must be submitted to BECC. Notice of the proposed project must be provided to affected communities, the sponsors must meet with affected groups, at least two public meetings must be held, and a community participation plan must be submitted to and approved by BECC, including a Comprehensive Community Participation Plan. Thus, these procedures allow for public participation in a process similar to that of a TEIA.

Public participation procedures are also inherent in a 1983 Agreement between the United States and the United Mexican States on Cooperation for the Protection and Improvement of the Environment in the Border Area (the La Paz Agreement), which requires an EIA when a project may have transboundary impacts. Although there are no specific references to public participation, impact assessments are to be undertaken in accordance with national legislation, and both the United States and Mexico have provisions for citizen involvement in EIA. It is important to note, however, that absent the inclusion of a non-discrimination requirement, it is unclear whether citizens in the United States or Mexico will be provided with opportunities to participate in the procedures taking place in the neighbouring country.

TEIA in Europe

EU Council Directives

Early in the evolution of TEIA principles and practice, European states – in part owing to geographical necessity and in part reflecting their growing political integration – were developing ways to address the challenges of conducting environmental assessments across national borders. In 1985, the European Communities adopted a Council Directive on the assessment of the effects of certain projects on the environment, which included some general provisions that potentially had transboundary application (EU Council Directive 1985).

In 1997, the Directive was amended to include more explicit provisions for TEIA (EU Council Directive 1997), and again in May 2003 to "contribute to the implementation of the obligations" arising under the 1998 UNECE Convention on Access to Information, Public Participation in Decision-making and Access to Justice in Environmental Matters (the Aarhus Convention) (EU Council Directive 2003). Among other provisions, the 2003 Directive: expands upon the information and timing requirements related to public participation in EIA, including notification requirements for interim and final decisions; explicitly requires parties to ensure "early and effective opportunities to participate" in EIA processes; entitles the public to "express comments and opinions when all options are open" to the authority taking the EIA-related decisions; and guarantees public access to a review procedure before a court or other impartial body to challenge the substantive or procedural legality of the implementation of the public participation requirements of the Directive.

Specifically related to TEIA, the Directive provides that the public of the affected state shall be notified no later than the public of the state of origin, and that domestic legislation shall provide them with the ability to participate effectively in the TEIA process.

The Helsinki Convention

The 1992 UNECE Convention on the Protection and Use of Transboundary Watercourses and International Lakes (the Helsinki Convention) focuses explicitly on the management of transboundary watercourses. The Convention contains provisions aimed at reducing and controlling transboundary impacts and calls upon State Parties to "develop, adopt, implement and, as far as possible, render compatible relevant legal, administrative, economic, financial and technical measures, in order to ensure, *inter alia*, that ... environmental impact assessment and other means of assessment are applied". Other related requirements include an innovative provision on joint monitoring and assessment, as well as requiring that the public be given access to relevant information.[6] The Convention also anticipates linkages with other relevant conventions, including the Espoo Convention, discussed below, governing transboundary EIA matters in the UNECE region.

Although the Helsinki Convention does not explicitly use the term "TEIA", with its intrinsic focus on transboundary watercourses and lakes, the Convention goes a long way toward establishing a legal and policy framework for TEIA to be used in the region's international watercourses. Moreover, in 1999, a Protocol to the Helsinki Convention on Water and Health was adopted explicitly in order to apply the provisions of access to information, public participation and access to justice under the Aarhus Convention to the specific context of international water management.

The Espoo Convention

The 1991 UNECE Convention on Environmental Impact Assessment in a Transboundary Context (the Espoo Convention) is arguably the most authoritative and specific international legal codification of TEIA.

The Convention requires member states to notify and consult each other on all major projects under consideration that are likely to have a significant adverse environmental impact across boundaries. The Appendices of the Convention enumerate a list of projects with transboundary effects requiring an EIA (upon which individual states may expand); outline procedural and content requirements for an EIA in a transboundary context; and provide guidance on which projects not categorically listed would trigger application of the Espoo Convention.

As a practical matter, the Espoo Convention requires that the country of origin open its EIA and decision-making procedures to the public and authorities in neighbouring, potentially affected states, taking their comments into account. Thus, the Convention utilizes a "non-discrimination" approach to participation, requiring that the public in a potentially affected country receive an opportunity to participate "equivalent" to that of the public in the country of origin. Should the party of origin have less stringent participation requirements and undertake responsibility for the notification and information exchange processes in a TEIA, the public in the affected country could receive fewer opportunities for participating in EIAs concerning transboundary impacts than for those whose impacts originate in their own country. If the affected country should undertake to conduct its own participation procedures, the time allotted for these processes by the country of origin may be inadequate for their more elaborate requirements. Similarly, should the country of origin have more specific participation requirements, the affected country is not required to implement those requirements in its own borders.

Even if the parties have similar requirements, the Convention does not define what constitutes "equivalent" participation opportunities. In order to participate in an equivalent manner, the public of affected countries may require translation of relevant documents, transportation and resources to attend public hearings within the country of origin, and perhaps additional time to that provided to the public in the country of origin because of the lag time in communications. All of these issues would need to be agreed upon between parties conducting a specific TEIA.

In the geographical context of Espoo, use of the non-discrimination principle is of less concern than it might be elsewhere, because most parties to the Convention are also signatories to the Aarhus Convention and the Helsinki Convention, and are members of the EU and thus subject to Directive 2003/35/EC. When taken together, these instruments provide strong minimum participation requirements and should lead to higher levels of stakeholder involvement in both national EIAs and TEIAs for shared watercourses throughout the region. Should Espoo be used as a model for other regions as they develop their own frameworks for

TEIA, however, the determination of minimum public participation requirements should be a priority to avoid confusion and even potential downward harmonization.

Espoo also places joint responsibility on the responsible authorities of the countries involved in TEIA for conducting public participation procedures. Parties to a specific TEIA must determine just *who* is responsible for notification and collection of comments, and what a "reasonable" time is for conducting these processes. If parties are not careful to specify, then notification might occur between responsible authorities and not reach the public of the affected party. Further, financing and translation for TEIAs are not addressed by the Espoo Convention, but can have a significant impact on the ability of the public in affected countries to participate meaningfully in the process.

To address the many questions raised by the Convention, in January 2004 the Secretariat of the Espoo Convention published a final draft "Guidance on Public Participation in Environmental Impact Assessment in a Transboundary Context", to be adopted by the working group on EIA (UNECE Secretariat 2004). This Guidance is discussed in detail below.

A Protocol on Strategic Environmental Assessment to the Espoo Convention (the SEA Protocol) was adopted in May 2003. The Protocol expands the commitments of the Convention to apply the principles of impact assessment to the preparation and adoption of plans, programmes, policies and legislation. The Protocol contains detailed provisions on public participation, including a broad statement that parties "shall ensure early, timely and effective opportunities for public participation, when all options are open, in the strategic environmental assessment of plans and programmes". This broadening of scope of impact assessment to decisions of more wide-ranging impact than just projects has the potential significantly to expand public access to and participation in environmental decision-making.

TEIA in practice

There is growing consensus that when a proposed project could have environmental effects on another nation a TEIA is necessary. As highlighted in the previous section, there is significant variability in the specificity of the existing legal requirements for TEIAs, which reflects the developing nature of this field and the degree of political integration of the regions in question. Practical experience with TEIAs is nascent as well, and what has occurred confirms significant variability in approach. Despite the dearth of available examples, there are some common elements of the TEIA process.

With regard to public participation, there is a clear consensus that public involvement is a critical and integral part of impact assessment. Stakeholder involvement has been an integral component in domestic EIAs since the process emerged in the US National Environmental Policy Act. Instruments that address TEIA have incorporated and expanded on this approach to include at least basic provisions for notification of the public of an affected country and for receiving the comments of those stakeholders at least once before the EIA is finalized.

In order to ensure that people who may be affected by a proposed project or activity have an opportunity to voice their concerns, international instruments usually promote either harmonization of procedures between states or non-discrimination, which is essentially an equitable safeguard to ensure that all affected people have equal opportunity to participate in environmental decision-making. The East African MOU, for example, promotes harmonization of EIA. The Espoo Convention and the draft North American TEIAA provide that participation must be non-discriminatory by prohibiting a state of origin from discriminating against neighbouring states and by mandating a domestic EIA system that complies with the Convention's minimum requirements – thereby facilitating harmonization as well as non-discrimination.

The most advanced examples of TEIA implementation are those originating in countries that are signatories to the Espoo Convention, as discussed below. Other examples of TEIA tend to be ad hoc, with co-operation driven by the issues that are the most economically, socially, environmentally and politically important to the states involved. In the absence of legally binding requirements, these particular circumstances often shape whether TEIA is necessary and what form it will take. The paucity of case studies assessing TEIA implementation – in particular, detailed accounts of the approaches taken to include the public in the assessments and how variations in those practices affect the environmental outcomes of TEIAs – suggests a need for further directed research in this area.

Implementation of the Espoo Convention

As noted above, the Espoo Secretariat recently responded to the parties' expressed need for guidance on public participation in the application of the Convention by developing Guidance based on a series of case studies (UNECE Secretariat 2004). The Guidance provides various means for resolving the ambiguities in the Convention's language with respect to timing, authority, translation and financial responsibility for participation, as well as what constitutes an "equivalent opportunity" for participation of the affected party's public. The Guidance also provides more general recommendations for parties to facilitate implementation of TEIA, including conducting preliminary work with potential TEIA participants, establishing points of contact with the public, and establishing bilateral and multilateral agreements and joint bodies for implementation.

Although the Guidance does not ultimately resolve the various issues surrounding the implementation of the Espoo Convention's public participation requirements, it does provide a clear articulation of the options and types of considerations that the parties should be addressing in implementing TEIA. For example, the Guidance points out the specific kinds of costs that may associated with public participation in a TEIA, provides examples from the case studies of how much was expended in a variety of circumstances, and outlines the various possibilities for who among the parties and the project proponent should assume these costs. The case studies on which the Guidance is based were not conducted utilizing a uniform methodology specifically aimed at assessing the effectiveness of the participation practices or facilitating evaluation of the project or policy outcomes. More focused research on TEIA implementation specifically structured to address such aspects of these and other transboundary assessments would be necessary to derive meaningful best practices for application in future TEIAs.

Nordic states

The Nordic states have a long history of international cooperation in environmental matters. As a result, it seems that the Espoo Convention complements an existing collaborative political framework. This framework began with the 1974 Nordic Convention on the Protection of the Environment, which allowed persons affected by nuisances caused by environmentally harmful activities originating in another state to bring proceedings challenging such activities in administrative tribunals or courts of the polluting state. This has avoided one of the major legal pitfalls encountered in transboundary cooperation to implement TEIA: how to overcome the issue of standing for non-nationals in environmental matters that directly affect them.

In 1996, the Nordic countries of Finland, Sweden, Denmark and Norway embarked upon a project called the Coordinated Application of the Espoo Convention (Tesli and Roar Husby 1999). One report from this project found that three main areas of the Convention relating to public participation required clarification in order for the coordinated implementation to be successful:
- designation of a responsible authority (as between the parties) for implementing participation;
- designation of methods of informing and soliciting comments from the affected parties' public; and

- which methods (as between parties) of participation should be utilized on both sides of the border.

Case studies undertaken for this study indicated that, without such clarification, participation by the affected parties' public tended to be ad hoc and take place only when the project proponent or a responsible authority took the initiative to ensure stakeholder involvement. The study's authors stressed that the Convention's allocation of shared responsibility for public participation resulted in a need for institutionalized reciprocity of participation methods, even with the Espoo Convention's non-discrimination principle in place.

The Baltic states

In the Baltic region, Estonia and Latvia concluded an agreement on implementation of the Espoo Convention in 1997. This agreement provides that the state of origin will bear the costs of any EIA and sets out the responsibilities of the parties for disseminating information. The Annex to this agreement includes a list of proposed activities within 15 km of the shared border that are subject to the agreement, which is more specific than the Espoo Convention. The agreement also establishes a commission, which decides, on a case-by-case basis, the procedural issues for conducting a TEIA, including the specific procedures for public participation. The establishment of a neutral, common body and the case-by-case approach set clear guidelines for the implementation of TEIA in Estonia and Latvia, thereby addressing the challenges indicated in the Nordic study.

These examples from Nordic and Baltic states show that specific regional and bilateral arrangements can help to clarify and enhance the coordination of the TEIA process. Although such regional and bilateral agreements may not necessarily be as effective in other regions, this approach appears to be working well in the European context, likely due at least in part to the additional legislative instruments (namely the Aarhus Convention and the EU Directives) that enhance the participatory and specific regulatory aspects of TEIA among member states. As the following examples articulate, however, legally binding arrangements in other regions of the world may not necessarily be the most effective means to advance TEIA practice.

Victoria Falls

The 1995 transboundary strategic environmental assessment for Victoria Falls provides an example of a successfully implemented TEIA with well-developed procedural guidelines. Victoria Falls is located on the border of Zambia and Zimbabwe on the Zambezi River and was declared a

World Heritage Site in 1989 by the United Nations Educational, Scientific and Cultural Organization (UNESCO 1989). The Zambian and Zimbabwean governments have a long history of cooperating to obtain the mutual benefits that Victoria Falls provides, particularly the economic benefits associated with tourism in the area. However, a four-fold increase in visitors during the period 1985–1995 in Zimbabwe, an increase in adventure tourism, and the need for additional infrastructure sparked concern over the potentially adverse environmental impacts associated with increased tourism and spurred the governments to assess the impacts and options for protecting the Falls (Silengo 1996).

The Zimbabwean and Zambian governments agreed to prepare a master plan for sustainable development in the Victoria Falls area to be implemented jointly by the two governments. To assist in implementing this plan, they decided that a transboundary strategic environmental assessment should be conducted to predict the cumulative environmental impacts of current and expected developments up to the year 2005, for an area within a 30 km radius of Victoria Falls. As noted above, SEA encompasses a broader assessment of longer-term and cumulative development changes indicated in new policies, programmes and plans, and attempts to make necessary recommendations based upon predicted impacts.

A comprehensive public consultation programme was organized, involving opinion surveys, workshops, "open houses" and media publicity. Some 150 stakeholders were involved in reviewing and commenting on the draft report and recommendations (Dalal-Clayton and Sadler 1996). Probably as a result of this extensive participation, the recommendations addressed issues of equitable benefits-sharing and poverty alleviation in local populations as an integral part of the plan.

The governments engaged the World Conservation Union (IUCN) – a neutral third party – to coordinate, direct and manage the TEIA and contributed themselves through a steering committee consisting of senior government officials from both states (Silengo 1996). The engagement of IUCN provided a politically neutral forum for dialogue and negotiation, and the organization was able to act as facilitator between the states and also to provide expertise in the various substantive areas of the study. The Canadian International Development Agency (CIDA) provided funding for the study, and the findings were utilized to prepare a skeleton management plan for the area as a contribution to the overall Master Plan.

Mexico and the United States

Historically, management of watercourses along the United States–Mexico border has not enjoyed particularly open communication or cooperation (Hayton and Utton 1989). However, there are recent indications that this may be improving.

An example of this improved cooperation and participation in TEIA between Mexico and the United States is the Tijuana and Playas de Rosarito Potable Water and Wastewater Master Plan. The Estuaries and Clean Waters Act of 2000 directs the United States Environmental Protection Agency (USEPA) to develop a comprehensive plan, with stakeholder involvement, to address transboundary sanitation problems in the San Diego–Tijuana border region. A significant component of this plan involves assessing the water and sanitation systems in the region, including the Colorado River, which flows across the border.

Three alternatives were formulated for the water system, and four alternatives were formulated for the sanitation system. The proposed Master Plan was followed by an environmental assessment, which was completed in February 2003 pursuant to NEPA and its implementing regulations. The environmental assessment analysed the potential environmental impacts, both local and transboundary, of the activities proposed in the draft Master Plan.[7] The Mexican environmental assessment also reviewed potential environmental impacts in Mexico. Transboundary effects were considered and analysed throughout the study.

This environmental assessment was subject to a 30-day public review period, during which the public and interested agencies from both nations were encouraged to submit comments. The EPA will consider all comments, including Mexican comments, on the environmental assessment as it finalizes the Master Plan. In this way, the San Diego–Tijuana border project exhibits a commitment to a more participatory TEIA process between the United States and Mexico.

Upper Mekong Navigation Improvement Project

In order to promote transportation along the Upper Mekong River, China, Myanmar, Laos and Thailand have proposed the Mekong River Navigation Improvement Project (Finlayson 2002). By removing 11 major rapids and 10 scattered reefs and shoals by "dredging and blasting", this project would "permit the passage of ships of 100–150 tonnes for 95% of the year" (Cocklin and Hain 2001: 6–7). A TEIA was prepared for the Mekong River Commission in September 2001. A TEIA team consisting of experts from China, Laos, Myanmar and Thailand initially went to 11 of the 21 working sites in order to produce a survey and collect hydrological data. The TEIA team found that there would be no long-term impacts on the fisheries and fishing-based livelihoods of communities along the Mekong River.

This TEIA was widely criticized as inadequate. The MRC commissioned independent evaluations of the TEIA, which have disputed the original assessment. They found that the proposed physical manipulations were intended to open the river to more traffic by larger ships and to expand economic activities, which might themselves introduce new pressures on the regional fisheries and fishing-based livelihoods. In addition, the extra pressures on the resources of the river and riparian lands could seriously affect the water quality (Finlayson 2002). The original TEIA, according to these independent analyses, is *"substantively inadequate* and in many places *fundamentally flawed"* (Cocklin and Hain 2001: 2).

The public participation elements for this TEIA are not known. This may be an example of public participation failing to be implemented, failing to be implemented so as to be meaningful, or failing to be used to correct a potentially flawed TEIA. This case study suggests there is a possible role for objective, independent assessments in TEIA, which may be advanced through regional institutions such as the MRC (which sought the independent expert evaluations) or through other organizations.

Key issues in the development of public participation in TEIA

Just as each river is unique yet shares certain features with other rivers, global TEIA processes and the role of public participation in them are characterized by both common and unique elements. A range of issues needs to be considered, including varying legal and regulatory structures, political systems, cultures, languages and specific socio-economic and environmental contexts. It is difficult to generalize a one-size-fits-all process for public participation in TEIA that is universally applicable. There are several areas nonetheless where experience with public participation mechanisms in TEIA suggests there needs to be additional consideration given as TEIA continues to develop.

Specificity and clarity of terms of agreement

The examples referred to throughout this chapter indicate that the most effective TEIAs have clear and specific terms of reference that states follow throughout the TEIA process, such as the bilateral agreement between Latvia and Estonia. Such specificity with respect to participation requirements increases transparency and helps ascertain responsibility, thereby avoiding disputes (or simple inaction) and facilitating implementation.

At present, there is a range of approaches, regulations and standards to protect against discrimination in participation across borders worldwide. TEIA processes that have incorporated a clear statement of terms have been able to operate more effectively and efficiently. In particular, it may help to establish, prior to initiating a TEIA, more rigorous planning and formalized specific requirements for public participation than is currently the practice. The Espoo Convention provides perhaps the clearest articulation of participation requirements but, as shown above, implementation of the Convention has highlighted the need for even more explicit terms.

Harmonization and non-discrimination

This chapter has made clear the value of harmonizing EIA participation procedures between states and of using the non-discrimination principle to ensure that all affected people have the opportunity to participate equally (Knox 2002). From the approaches taken worldwide, there appears to be a consensus that, at a minimum, the originating state should accord the same protections and access to information to the public of affected states as to individuals within its own borders (UNECE Espoo Convention 1991; Aarhus Convention 1998; CEC 1997).

The recent revisions to the International Law Association (ILA) Rules on Equitable and Sustainable Use of Waters, which are reflective of customary international law, recognize the duty of the state to take reasonable steps in the management of waters to ensure that persons affected by those decisions are able to participate in the processes through which those decisions are made (ILA 2003). Water-body-specific and regional instruments (such as the Helsinki Convention and the East African MOU) also require states to ensure the involvement of the affected states' populace, and there is widespread practice of nations pushing for increasingly participatory impact assessment processes (e.g. at the national level through EIA). This coalescing of state practices and international norms creates a strong foundation upon which TEIA public participation frameworks are likely to grow.

Formalizing TEIA practices

Many TEIA procedures, including many of the examples discussed in this chapter, tend to be in the form of framework and other more general provisions that frequently are non-binding. Although the Espoo Convention and the evolving EAC Environmental Assessment Guidelines for Shared Ecosystems in East Africa (EAC 2004) are exceptions, TEIA and its participation procedures remain in the nascent stages of being formalized into binding regional or international agreements.

There are a number of possible reasons why this may be the case. First, the principles of TEIA are still evolving, and reaching consensus regionally (let alone globally) on the details has proven to be challenging. Second, formalizing the principles of non-discrimination and harmonization would require states to give citizens of another state the right to access their legal processes, which remains a controversial step. Thus, maintaining TEIA at a more ad hoc level may be more politically palatable for many states. Third, a non-binding process may facilitate cooperation and dialogue, advancing and refining approaches to TEIA more rapidly and more specifically than would a legally binding treaty-making process. A non-binding approach is more flexible in granting discretion to states with respect to when and how to conduct a TEIA and therefore is perhaps more likely to be adopted. Legally binding arrangements, on the other hand, are stricter in form and mandate principles to which states are legally bound (Gray 2000).

However, failure to codify TEIA and its participation procedures as binding legal instruments may mean that TEIA becomes a tool of political convenience instead of one of environmental discipline. Allowing TEIA to be an ad hoc procedure also leaves the public participation provisions at the discretion of the originating state. Allowing states discretion in the public participation provisions of any assessment process leaves room for processes that involve the public in name only or that result in neither state taking responsibility for participation. Resting TEIA upon customary law as opposed to a more formalized system may, therefore, undercut the usefulness, reliability and inclusion of TEIA and its public participation mechanisms.

Role for non-state actors

Existing regional associations such as the European Union, the East African Community, the International Joint Commission and the North American Commission on Environmental Cooperation have facilitated cooperation, enabling a common institutional structure to operate within the region. Such regional associations, which have experience with cooperation and coordination, can often facilitate consensus.

River basin management authorities, such as the Mekong River Commission, provide another example of non-state actors that have, in many instances, conducted TEIA for many years in practice if not in name. As these authorities continue to spread and cross borders, they may become the agents and beneficiaries of improved TEIA participation processes.

NGOs also have an increasing role to play by providing an objective,

> **"Parallel" public participation**
>
> In the Czech Republic, NGOs reacted to what they perceived as a failure to include the public early and comprehensively enough to enable meaningful participation. They responded by holding special public meetings in potentially affected communities to inform the local population about the formal EIA process and documentation. These organizations encouraged the public to submit their comments, gathered those submitted and submitted them to the competent authority in the appropriate format for review. Such "parallel public participation" points to a potential role for, *inter alia*, existing local and regional water management institutions in facilitating the further elaboration and implementation of TEIA in the context of specific watersheds. (Richardson et al. 1998: 201)

subjective or impartial voice, depending on the particular NGO. As such, they can act as mediators, facilitators, overseers, advocates and activists as well as credible sources of information, as evidenced by the work of the IUCN in the Victoria Falls SEA.

Additionally, states are sometimes incapable of providing meaningful opportunities for the public to participate in TEIA because the necessary legal framework is not in place, the resources are unavailable or states are reluctant to open such processes to stakeholders. Where this is the case, NGOs and river basin organizations can play a key role in ensuring that the public can access and contribute to impact assessments.

Financial resources for meaningful public participation

Involving the public in an effective and meaningful way requires an investment, sometimes significant, of time and resources. In places where resources for such processes are limited, as is often the case for projects taking place in many developing countries or countries in transition, the technical aspects of the assessment often take priority over public consultation. This is partially owing to the fact that project proponents or relevant government agencies often overlook the value of public input and merely perceive the public as lacking the necessary technical or scientific knowledge to contribute meaningfully to an assessment. Unless the public are involved in a thoughtful way, the public participation element may fail to educate the public about the project or allow useful comment, thus undermining local people's political will.

In a developing country context, funding is often lacking to conduct a comprehensive TEIA, and public participation can take a backseat to the more technical aspects of the process when limited resources must be allocated. The World Bank, the regional development banks and bilateral institutions (such as the Canadian International Development Agency in the Victoria Falls SEA) can be essential in meeting this need. Such institutions can supply much-needed funding and, equally importantly at times, expertise. They can also help to ensure that certain TEIA procedures are followed if such procedures do not otherwise exist or, at times, even if they do. This exchange of funding, experience and expertise can be crucial – and has proved to be so – in developing and implementing TEIA.

The secretariats of the regional agreements may, at times, serve a similar function, as a source of either funding, expertise or accountability, or all of the above. Organizations such as the New Partnership for Africa's Development and others may come to play similar roles.

Dispute resolution and access to justice

Following publication of a final TEIA, citizens, governments, institutions and organizations may seek an avenue through which to appeal against an unsatisfactory analysis or decision. In most cases, these avenues are still limited or lacking. To the extent that there is public access to dispute resolution, it is usually through national courts, although constitutional or legal impediments may preclude members of the public in the affected state from bringing an action in the originating state.

International organizations such as the World Bank and other regional development banks increasingly provide internal administrative mechanisms for dispute resolution, such as inspection panels (Bernasconi-Osterwalder and Hunter 2002). Access to these quasi-judicial mechanisms is usually predicated on an alleged failure to follow the institution's internal policies or procedures, such as those governing environmental assessment.

There remain significant opportunities for regional organizations such as river basin organizations and NGOs to mediate disputes, especially in an informal way. The Upper Mekong Navigation Improvement Project described above demonstrates the important role that the MRC played in providing such an avenue of appeal for aggrieved states. It was, after all, the MRC that took on Lao PDR's complaints that the initial EIA conducted for the Upper Mekong Navigation Improvement Project was inadequate and referred the EIA to independent experts. Still, much remains to be done to provide for effective dispute resolution between states in the management of international watercourses, let alone to ensure public access to such mechanisms.

Conclusion

This chapter has traced the evolution of TEIA from its roots in EIA to its inclusion in international agreements, customary law and other instruments. TEIA represents a practical vehicle for implementing Principle 10 of the Rio Declaration on Environment and Development (UNCED 1992) through its emphasis on access to information, public participation, harmonization and non-discrimination. TEIA, although crystallizing in different regions throughout the world, is still in the formative stages.

TEIA is particularly important to management of international watercourses. Water is likely to become an increasingly critical issue in coming decades. Transboundary rivers and lakes pose a particular challenge owing to the political, economic and cultural coordination that is required to manage water adequately among states and the potential for conflict between states. TEIA has the potential to help mitigate the management difficulties associated with increasing water scarcity through a transparent, participatory and deliberate decision-making process. The public participation provisions of TEIA in particular may help avoid conflict by ensuring that the citizens of both affected and originating states have a say in the management of the water resource.

As experience is gained with TEIA implementation, there is also potential to expand its scope and apply its lessons not only to environmental impacts but also to social, cultural, health and other related impacts. As with the environmental assessment, this could be done not only to assess proposed projects, but also at the strategic level of plans, programmes and policies. Such coordination could also help mitigate future conflicts.

The development of TEIA will most likely be driven by past example and present local conditions. As more TEIAs are undertaken, important experience in implementing them and their public participation methods will be gained and should be captured. Thus, there is a distinct need for further articulation and dissemination of case studies, not only within but also between regions, to develop "lessons learned" in TEIA and transboundary SEA implementation. Pilot projects that focus specifically on assessing the effectiveness of participation practices and the role such practices play in project and policy outcomes could provide meaningful insight into best practices. In the end, this will enable more effective and equitable environmental management of shared watercourses.

Notes

1. For example, the US National Environmental Protection Act requires the government agency conducting the assessment to publish a "notice of intent" in the *Federal Register* describing the proposed action, stating whether any scoping meetings will be held, and providing the name and address of a person within the agency who can answer questions and receive comments on the proposal (Code of Federal Regulations). EIA legislation in Denmark and in the Netherlands provides that a decision not to proceed with an impact assessment must be made public and subject to comment and review (Stærdahl et al. 2003); and Bulgaria's Environmental Protection Act contains an innovative provision whereby concerned persons are authorized to submit EIA proposals for activities they believe should be covered by the process (Teel 2001).
2. Public participation also provides several benefits for those stakeholders who contribute to the EIA process. In order to provide meaningful input, the public must be given access to the relevant information regarding project proposals and their potential impacts. This transparency engenders increased accountability on the part of decision makers, ensuring that conclusions are reasoned and defensible and can build trust and encourage further cooperation between the public and the authorities responsible for overseeing the EIA process. Increased access to information can also improve public understanding of how decision-making processes work, which in turn can create a greater sense of empowerment and social responsibility.
3. For a detailed overview of the legal foundations of TEIA in international law, see Cassar and Bruch (2004).
4. See also Resources Policy Support Initiative, Program on Mekong Regional Environmental Governance, available at http://www.ref-msea.org/mreg.html.
5. In particular, Article XIV(2)(b) requires parties to "ensure that policies, plans, programmes, strategies, projects and activities likely to affect natural resources, ecosystems and the environment in general are the subject of adequate impact assessment at the earliest possible stage and that regular environmental monitoring and audit are conducted". Article XVI also requires the States Parties to adopt the legislative and regulatory mechanisms necessary to ensure access to information, participation in decision-making, and access to justice in the context of impact assessment.
6. "Riparian Parties shall ensure that information on the conditions of transboundary waters, measures taken or planned to be taken to prevent, control and reduce transboundary impact, and the effectiveness of those measures, is made available to the public." Article 16 of the Helsinki Convention further highlights the importance of reasonable timeframes and access that is free of charge. The parties "shall ensure that this information shall be available to the public at all reasonable times for inspection free of charge, and shall provide members of the public with reasonable facilities for obtaining from the Riparian Parties, on payment of reasonable charges, copies of such information".
7. Assessments were conducted for, among other things: air quality, surface water, groundwater, biological resources, cultural resources and noise (USEPA 2003).

BIBLIOGRAPHY

Algiers Convention (2003) *African Convention on the Conservation of Nature and Natural Resources (Revised Version)*, opened for signature 11 July 2003.

BECC [Border Environment Cooperation Commission] (1993) *Agreement between the Government of the United States of America and the Government of the United Mexican States Concerning the Establishment of a Border Environment Cooperation Commission and a North American Development Bank.*
Bernasconi-Osterwalder, F. and D. Hunter (2002) "Democratizing Multilateral Development Banks", in C. Bruch (ed.), *The New "Public": The Globalization of Public Participation*, pp. 95–103. Washington, DC: Environmental Law Institute.
Boundary Waters Treaty (1909) *Treaty Relating to Boundary Waters between the United States and Canada, Jan. 11, 1909*, U.S.-Gr. Brit., 36 Stat. 2448.
Bruch, C. (2003a) "African Environmental Governance: Opportunities at the Regional, Subregional and National Levels", in B. Chaytor and K. Gray (eds), *International Environmental Law and Policy in Africa*, p. 217. Dordrecht: Kluwer Academic Publishers.
——— (2003b) "Role of Public Participation and Access to Information in the Management of Transboundary Watercourses", in M. Nakayama (ed.), *International Waters in Southern Africa*. Tokyo: United Nations University Press.
Cassar, A. and C. Bruch (2004) "Transboundary Environmental Impact Assessment in International Watercourse Management", *New York University Environmental Law Journal* 12: 169.
CEC [North American Commission on Environmental Cooperation] (1997) *Draft North American Agreement on Transboundary Environmental Impact Assessment*, available from ⟨http://www.cec.org/pubs_info_resources/Law_treat_agree/pbl.cfm?varlang=english⟩.
Cocklin, C. and M. Hain (2001) *MRC, Evaluation of the EIA for the Proposed Upper Mekong Navigation Improvement Project 2*, http://www.irn.org/programs/mekong/021018.socialimpacts.pdf (citing Joint Experts Group on EIA of China, Laos, Myanmar, and Thailand, *Report on Environmental Impact Assessment: The Navigation Channel Improvement Project of the Lancang-Mekong River from China-Myanmar Boundary Marker 243 to Ban Houei Sai of Laos*, 2001).
Dalal-Clayton, B. and B. Sadler (1996) *The Status and Potential of Strategic Environmental Assessment: Case 6: SEA of Development around Victoria Falls*, available from ⟨http://www.iied.org/spa/sea.html#case⟩.
EAC [East African Community] (1998) *Memorandum of Understanding between the Republic of Kenya and the United Republic of Tanzania and the Republic of Uganda for Cooperation on Environment Management*, 22 October 1998.
——— (1999) *Treaty for the Establishment of the East African Community, Nov. 30, 1999, African Yearbook of International Law* 7: 421 (entered into force 7 July 2000).
——— (2004) *Environmental Assessment Guidelines for Shared Ecosystems in East Africa*, signed December 2004 (on file with authors).
Estonia–Latvia Agreement (1997) *Agreement on Environmental Impact Assessment in a Transboundary Context, Mar. 14, 1997*, Est.-Lat., 1986 U.N.T.S. 116.
European Bank for Reconstruction and Development (2003) *Environmental Policy*, available from http://www.ebrd.com/about/policies/enviro/policy/policy.pdf.

EU Council Directive (1985) 85/337/EEC, 1985 O.J. (L 175).
—— (1997) 97/11/EC, 1997 O.J. (L 073).
—— (2003) 2003/35/EC, 2003 O.J. (L 156).
Finlayson, B. (2002) *Report to the Mekong River Commission on the "Report on Environmental Impact Assessment: The Navigation Improvement Project of the Lancang-Mekong River from China-Myanmar Boundary Marker 243 to Ban Houei Sai of Laos"*, 2, University of Melbourne, available from ⟨http://www.irn.org/programs/mekong/021018.critiquehydrology.pdf⟩.
Gray, K. (2000) "International Environmental Impact Assessment: Potential for a Multilateral Environmental Agreement", *Colorado Journal of International Environmental Law and Policy* 11: 83.
Hayton, R. and A. Utton (1989) "Transboundary Groundwaters: The Bellagio Draft Treaty", *Natural Resources Journal* 29: 663.
Hildén, M. and E. Furman (2002) "Towards Good Practices for Public Participation in the Asia-Europe Meeting Process", in C. Bruch (ed.), *The New "Public": The Globalization of Public Participation*. Washington, DC: Environmental Law Institute.
Hughes, R. (1998) "Environmental Impact Assessment and Stakeholder Involvement", in A. Donnelly, B. Dalal-Clayton and R. Hughes (eds), *A Directory of Impact Assessment Guidelines*. Nottingham, UK: Russell Press.
ILA [International Law Association] (2003) *The [Revised] International Law Association Rules on Equitable and Sustainable Use of Waters*, 10th draft.
Knox, J. (2002) "The Myth and Reality of Transboundary Environmental Impact Assessment", *American Journal of International Law* 96: 291.
La Paz Agreement (1983) *Agreement between the United States of America and the United Mexican States on Cooperation for the Protection and Improvement of the Environment in the Border Area, Aug. 14, 1983*, U.S.-Mex., art. 7, 35 U.S.T. 2916, 2919, 22 I.L.M. 1025, 1027–28.
Moreno, I., et al. (1999) "Free Trade and the Environment: The NAFTA, the NAAEC, and Implications for the Future", *Tulane Environmental Law Journal* 12: 405.
MRC [Mekong River Commission] (1995) *Agreement on the Cooperation for the Sustainable Development of the Mekong River Basin*, 5 April, 34 I.L.M. 864.
—— (2002a) *Annual Report 2001*, p. 10, available from ⟨http://www.mrcmekong.org/pdf/annual_report_2001.pdf⟩.
—— (2002b) *Annual Report 2002*, p. 16, available from ⟨http://www.mrcmekong.org/pdf/annual_report_2002.zip⟩.
—— (n.d.) *Public Participation in the Context of the MRC*, available at ⟨http://www.mrcmekong.org/document_online/document_online.htm⟩.
NAAEC (1993) *North American Agreement on Environmental Cooperation, Sept. 14, 1993*, 32 I.L.M. 1480, entered into force 1 January 1994.
Nicro, S., et al. (2002) "Public Involvement in Environmental Issues: Legislation, Initiative and Practice in Asian Members of ASEM Countries", in Thailand Environmental Institute, *Public Involvement in Environmental Issues in the ASEM – Background and Overview*, pp. 37–40, available from ⟨http://www.vyh.fi/eng/intcoop/regional/asian/asem/asem.pdf⟩.

Nordic Environmental Protection Convention (1974) *Convention on the Protection of the Environment between Denmark, Finland, Norway and Sweden, Feb. 19, 1974*, Art. 3, 1092 U.N.T.S 279, 296 (1978).

OECC [Overseas Environmental Cooperation Center, Japan] (2000) *Environmental Impact Assessment for International Cooperation*, §2.1, available from ⟨http://www.env.go.jp/earth/coop/coop/materials/10-eiae/contents.html⟩.

Paisley, R. (2002) "Adversaries into Partners: International Water Law and the Equitable Sharing of Downstream Benefits", *Melbourne Journal of International Law* 3: 280.

Preiss, E. (1999) "The International Obligation to Conduct an Environmental Impact Assessment: The ICJ Case Concerning the Gabcikovo-Nagymaros Project", *New York University Environmental Law Journal* 7: 307.

Richardson, T. et al. (1998) "Parallel Public Participation: An Answer to Inertia in Decisionmaking", *Environmental Impact Assessment* 18: 201.

SADC [Southern African Development Community] (1992) *Treaty of the Southern African Development Community, Aug. 17, 1992*; reprinted in 32 I.L.M. 116 (1993).

―――― (2001) *Revised Protocol on Shared Watercourses in the Southern African Development Community (SADC), Aug. 7, 2000*, 40 I.L.M. 321.

Scanlon, J., A. Cassar and N. Nemes (2003) "Water as a Human Right?", paper prepared for the 7th International Conference on International Environmental Law: Water and the Web of Life, available from ⟨http://www.iucn.org/themes/law/pdfdocuments/WW-Rev%202%20-%202nd%20June.pdf⟩.

SEA Protocol (2003) *Protocol on Strategic Environmental Assessment to the Convention on Environmental Impact Assessment in a Transboundary Context, adopted May 21, 2003*, available from ⟨http://www.unece.org/env/eia/documents/protocolenglish.pdf⟩.

Sikoyo, G. (2005) "Public Participation in the Development of Guidelines for Regional Environmental Impact Assessment (EIA) of Transboundary Aquatic Ecosystems of East Africa", in C. Bruch et al. (eds), *Public Participation in the Management of International Freshwater Resources*. Tokyo: United Nations University Press.

Silengo, M. (1996) "SEA of Developments around Victoria Falls, Zambia", in B. Sadler, *International Study of the Effectiveness of Environmental Assessment*, for the Canadian Environmental Assessment Agency and International Association for Impact Assessment. Minister of Supply and Services, Canada.

Stærdahl, N., et al. (2003) *Environmental Impact Assessment in Thailand, South Africa, Malaysia and Denmark*, Working Report of the Critical Comparative EIA in 4 Countries Network, available from ⟨http://www.ruc.dk/teksam/omteksam/Kortom/international/workingpaper1/⟩.

TAI [The Access Initiative] (n.d.) at ⟨http://www.accessinitiative.org⟩ (accessed 4 December 2003).

Teel, J. (2001) "International Environmental Impact Assessment: A Case Study in Implementation", *Environmental Law Reporter* 31.

Tesli, A. and S. Roar Husby (1999) "EIA in a Transboundary Context: Principles and Challenges for a Coordinated Nordic Application of the Espoo Convention", *Environmental Impact Assessment Review* 19: 57.

Tijuana River Valley Estuary and Beach Sewage Cleanup Act of 2000, Pub. L. No. 106–457, §§801–06, 114 Stat. 1957, 1977–81 (2000).

Timoshenko, A. (1988) "The Problem of Preventing Damage to the Environment in National and International Law: Impact Assessment and International Consultations", *Pace Environmental Law Review* 5: 475.

UNCED [United Nations Conference on Environment and Development] (1992) *Rio Declaration on Environment and Development*, UN Doc. A/CONF.151/5/Rev.1, reprinted in 31 I.L.M. 874.

UNECE [United Nations Economic Commission for Europe] (1991) *Convention on Environmental Impact Assessment in a Transboundary Context, Feb. 25, 1991* [Espoo Convention], 1988 U.N.T.S. 310 (1997).

—— (1992) *Convention on the Protection and Use of Transboundary Watercourses and International Lakes, Mar. 17, 1992* [Helsinki Convention]; reprinted in 31 I.L.M. 1312.

—— (1998) *Convention on Access to Information, Public Participation in Decision-Making and Access to Justice in Environmental Matters, June 25, 1998* [Aarhus Convention]; reprinted in 38 I.L.M. 517, entered into force 30 October 2001.

UNECE Secretariat (2004) *Guidance on Public Participation in Environmental Impact Assessment in a Transboundary Context*, available from ⟨http://www.unece.org/env/eia/documents/draft%20decisions⟩/Decision%2008%20-%20guide%20-%20English.pdf⟩.

UNEP [United Nations Environment Programme] (1987) *Governing Council Decision: Goals and Principles of Environmental Impact Assessment*, princ. 4, UNEP/GC.14/17 Annex III, UNEP/GC/DEC/14/25; reprinted in UNEP (1987), *Principles of Environmental Impact Assessment, Environmental Policy and Law* 17: 36.

—— (2002) *Environmental Impact Assessment Training Resource Manual* 112, 2nd edn, available from ⟨http://www.unep.ch/etu/publications/EIAMan_2edition_toc.htm⟩.

UNESCO [UN Educational, Scientific and Cultural Organization] (1989) *Report of the World Heritage Committee*, 13th Sess. at 13, UN Doc. SC-89/CONF.004/12.

USEPA [US Environmental Protection Agency] (2003) *Environmental Assessment: Tijuana and Playas de Rosarito Potable Water and Wastewater Master Plan*, sec. 2.1, at 4-1-4-11.

WHO [World Health Organization] (2004) *Water, Sanitation and Hygiene Links to Health: Facts and Figures*, available from ⟨http://www.who.int/docstore/water_sanitation_health/General/facts&fig.pdf⟩ (last updated November 2004).

World Bank Group (2004) *Millennium Development Goals*, available from ⟨http://www.developmentgoals.org/⟩ (last updated November 2004).

Part II
"Information technology" approaches

5

The Internet and e-inclusion: Promoting on-line public participation

Hans van Ginkel and Brendan Barrett

Introduction

James Madison, the fourth President of the United States, argued that "a popular Government without popular information, or the means of acquiring it, is but a Prologue to a Farce or Tragedy, or perhaps both" (Browning 1996: 1). Over 170 years later, the same notion holds true as we increasingly see information and communication technologies being used to gather and analyse the data that governments need. Looking even further back, it was Thomas Jefferson, the third President of the United States, who argued that, "whenever the people are well-informed, they can be trusted with their own government. Whenever things get so far wrong as to attract their notice, they may be relied on to set them to rights."[1] In this context, when reviewing the implications of e-government for the digital divide and information-sharing, Riley (2004) states:

Governments have traditionally distributed wide amounts of information to citizens to ensure the execution and administration of government programs. All areas of society receive some form of minimal information from government whether it is essential facts for tax filing, weather information for the public, trade data for businesses, statistical studies, or job opportunities. The public is used to being informed through advertisements in all communications media, television, radio, the Internet, newspapers, magazines, pamphlets, brochures, billboards, or whatever medium is best to get the message out. Governments write millions of

83

words in reports and studies and make them available to particular segments of society that have a need for the knowledge.

The above argumentation suggests that governments need information from citizens in order to govern and that citizens need information in order to participate effectively. We now have a new term for this in the information age: e-democracy. According to Riley (2004):

> The term electronic democracy no longer refers to simply the involvement in the political process or being able to interact with the government officials or participate in online consultations. The ability of individuals to share information and knowledge amongst themselves has now come under the rubric of e-democracy. Such sharing is an extension of what has occurred for centuries between peoples, groups and governments. Facilitating information sharing through the use of information and communication technologies is as much a duty of government as it is a practice in democracy by the citizen.

This is the foundation of deliberative e-democracy – an unwritten contract between the governed and the governing. This democratic contract has been tested on many occasions, most dramatically in the more turbulent times of civic protest in the 1960s and 1970s across the globe. In the midst of this period, Sherry Arnstein introduced the 10-step ladder of participatory democracy when she identified the provision of information as the most important first step to legitimate participation (Arnstein 1969).[2]

In the Information Society, however, we find yet again that the contract is being tested. Now we are more concerned with problems of information overload (Hill and Hughes 1998)[3] or exclusion by information (large amounts of information are available on individuals and can be utilized in discriminatory fashion – see Perri 6 and Jupp 2001). We live in the personal information economy where personalization rather than mass production of information is key (Negroponte 1995; Perri 6 and Jupp 2001). The situation may be getting worse not better, and we have new expressions that capture the essence of these problems, such as "Information Fatigue Syndrome". Another often heard term is "data smog", which relates to the existence of too much low-quality information (Schenk 1998).

Risk, environmental information and the Internet

It is interesting that in the context of information overload we would find ourselves using an environmental term to describe the problem. This brings us to the second cause for concern. We now find ourselves dealing with the increasingly complex nature of environmental problems and risks at both the global scale (for example, climate change, loss of biodiversity) and the local scale (e.g. toxic pollutants, endocrine disruptors). What impact does this complexity have on society? Faced with complex environmental issues and ever-growing information flows, the central question becomes – can democracy and public participation flourish in today's complex technological information society?

In his book on *Citizens, Experts, and the Environment*, Frank Fischer (2000) argues that the Information Society as an ideology presents technological advance as social progress and it conflates the concepts of information and knowledge. Reading between the lines, Fischer appears to recommend that we, as citizens, reflect upon whether we are really witnessing societal progress through the use of information technology (IT) or just the rapid development of some form of mass distraction. In other words, the overwhelming flows of information on the Internet might present a real danger of people becoming disengaged from existing political processes and simply using these tools for their entertainment value – lost in cyberspace. As Sherry Turkle reminds us, the computer and the Internet are new mediums to "project our ideas and our fantasies" (1995: 9).

Referring specifically to the question of environmental risk, Fischer (2000: 4) warns that,

as the growing influence of science and technology gives rise to increasingly public fears and disputes about its privileged status, laypersons express political uncertainty and hesitation about the implementation of scientific and technological projects – from nuclear energy to biotechnology. More and more environmental groups, citizens, and politicians speak of the need to regulate and control science. While the scientific community complains of intervention in the pursuit of knowledge, the public increasingly comes to see that scientists are themselves lay persons in matters concerning political goals and social judgements. Bringing these scientific and normative judgements together requires new institutional forums.

Some commentators see the Internet as a powerful forum for dialogue on environmental issues, although still at the experimental stage (Beierle and Chahill 2000; Barrett et al. 2001). Perhaps more dramatically, others claim that the Internet is the greatest hegemonic device ever created by humankind and that it will lead increasingly to a globally monolithic, monocultural and technocratic world (Bowers 2000). This homogeneous form of modern society would run counter to our environmental and cultural needs and to the preferences, as expressed by many including the Deep Ecologists and other environmental ideologies, to retain or recreate small-scale and autonomous communities (Devall and Sessions 1986;

Katz et al. 2000; Shumacher 1973). Technological progress is driving forward a new form of what Bauman (2000) calls "liquid modernity", where our notions of everything (identity, community, time/space, etc.) are fluid, continuously and irrecoverably changing. Taking the above perspectives into consideration, we find ourselves torn between the perception of the Internet as the panacea for contemporary social and environmental problems or a potential Pandora's box containing unimaginable woes ready to be unleashed on an unprepared world.

Both viewpoints illustrate the multifaceted nature of notions such as e-governance and the need to critically evaluate the implications that the Information Society would have for participatory forms of governance. We need to ask whether the use of the Internet contributes to the development of discursive institutions capable of rapid reaction to the stresses and pressures, particularly in the environmental arena, or whether it will turn into more of a tool for manipulation and control? The potential positive and negative ramifications of the Internet for wider society have been extensively documented (Slevin 2000; Mitchell 2000; Toregas 2001). Nevertheless, it is only recently that researchers have focused on the possible implications for community engagement within the framework of emerging forms of digital or e-governance (Perri 6, 2004). Quite clearly some kind of IT-led transformative process is under way with the potential to alter the modus operandi of interaction between governmental bodies and the general citizenry. Nevertheless, little is known about the direction of current changes and their potential implications for the future forms of governance.

The Internet and inclusive governance – The e-inclusion debate

From the wide range of thematic areas associated with the Information Society and its potential to influence development (i.e. e-democracy, e-government, tele-education, e-commerce, tele-services, telework, digital divide and social exclusion), it is important in this context to focus narrowly on the scope for web-based interactive forms of civic engagement. Although we recognize that there are many significant barriers to the adoption of information technologies for public participation purposes, some clear ideas are emerging on their potential application in order to attempt to bridge the contemporary "perception divide" between governments and the communities they serve.

What is the perception divide? This term describes the situation where the administration (national and local politicians, officials, experts) are not on the same wavelength as their community for a variety of reasons, such as a lack of communication, the tendency to form expert cliques or just plain arrogance along the lines of "we know best". Let us now look at some of the issues associated with the use of the Internet in public participation in more detail.

Perhaps the first point to note is that globally we are moving slowly but surely towards the first billion people on-line (the latest figures indicate around 900 million on-line in 2005[4]). This represents around 14 per cent of the world's population. The remainder of the world's population is currently on the wrong side of the digital divide. Furthermore, it is essential to highlight that, within the Internet "connected", another divide exists. The total number of broadband lines at the end of 2004 was around 150 million (data transfer speeds of 256 kbit/s and greater are commonly marketed as "broadband") (Point Topic 2005). We are now witnessing various layers of connectivity to the Internet – from narrow to broadband – as well as variations spatially and according to economic status. For instance, a survey of Internet connectivity in Japan back in 1999 revealed some regions where less than 3 per cent of the population was on-line (reported in Barrett and Yamada 2000). Although the situation in Japan has improved in the meantime, a more recent survey in England and Wales revealed significant variations in terms of access to broadband, as shown in Figure 5.1. This distinction is very important. A 2004 report from the University of Southern California on Internet use in the USA confirms that: "Compared to modem users, broadband users spend more hours online working on their jobs at home, instant messaging, playing games, seeking entertainment information, using online auctions, and downloading music." Moreover, "Compared to broadband users, modem users spend more hours online reading e-mail, seeking information on hobbies, Web browsing, schoolwork, and looking for medical information" (USC Annenberg School Center for the Digital Future 2004: 37).

As explained by Riley (2004), when looking at the question of the digital divide and Internet connectivity it is clear that "certain elements of society, the economically underprivileged, the illiterate, disabled, or disenfranchised, might fail to reap the benefits of e-government services". This situation is further compounded when comparing developed and developing countries. Although access to computers and Internet connections are important start points in bridging this divide, some commentators also stress the need to address issues associated with on-line content – which in itself may prove exclusive. Perri 6 and Ben Jupp (2001), in particular, highlighted four aspects of the content problem:

(1) the high costs of access to content on-line may prove insurmountable for some groups in society;
(2) paternalistic forms of content targeted at specific groups in society by

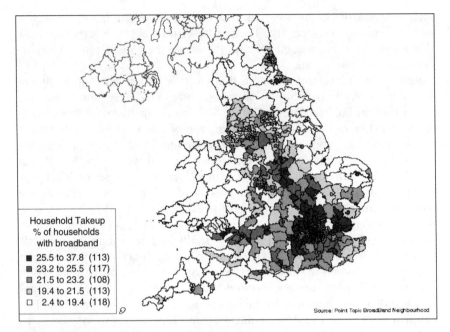

Figure 5.1 Access to broadband in England and Wales, 2004.
Source: Point Topic Broadband Neighbourhood, ⟨http://www.point-topic.com/home/bbn/⟩.

mainly public sector entities are unable actually to reach those groups owing to the costs associated with production or an inability to find space on existing media or to compete with other diverse forms of content;
(3) the costs of production prohibit socially excluded people from creating their own content; and
(4) there is an increased prevalence of bigoted forms of content.

This leads us to conclude that the ideal of a content-rich Internet comes with strings attached.

In response to these issues related to the digital divide, we have seen the emergence of a new social movement rallying around the term "e-inclusion", particularly in Europe. This is a central plank in the European Commission's strategy for e-Europe under the slogan – *An Information Society for All* (CEC 2000). E-inclusion is seen as the ability to close the gap between developed and less developed countries, and within countries; to promote democracy and mutual understanding; and to empower disadvantaged individuals, such as the poor, the disabled and the unemployed. This is quite a lot to ask from one social movement! The contemporary "digital divide" at the global level is clear to all observers, but what remains uncertain is the potential impact on the distribution of power, wealth, privileges and freedoms in all corners of the world that the Internet could bring. Social projects that seek to bridge the digital divide by providing greater community access to information technologies, although fundamentally important per se, must be critically scrutinized in the context of the motivations of the stakeholders involved in project promotion.

Information technologies and public participation in environmental governance

There are two potential implications of widespread use of information technologies to support public participation. First, there is the information-processing and dissemination element, where we see increasingly sophisticated environmental information in diverse forms (including via geographical information systems) disseminated in real time via the Internet. Second, we see the emergence of new forms of civic engagement through websites that promote on-line interaction between citizens and government policy makers with the use of a range of tools (Hill and Hughes 1998). For instance, a study by the British Council published in 1999 supports this supposition (British Council 1999). This study looked at emerging practice with the application of the Internet to public participation and indicated five possible benefits, as follows:
- increased information accessibility;
- greater public involvement;
- public awareness-raising;
- promotion of enhanced communication;
- stimulation of discussion on the merits of e-governance.

Moreover, the British Council study argued that e-governance can be defined as encompassing the use of a variety of information technology tools by government in order to connect directly with citizens and to enhance service delivery, provide for sustainable economic development and safeguard democracy.

Another recent review of the experience and potential use of e-governance to support development across the globe outlined the main benefits of Internet use in terms of cost reductions, producing more for less, and achieving results more rapidly, to a higher quality and in new ways (Heeks 2001). Nevertheless, the same study identified six barriers hindering the degree of "e-readiness" of countries in different parts of the world. These are basically infrastructure problems associated with data systems (i.e. the quality of data and their security), regulations, institutions, human capacities, technology and leadership (i.e. the existence or the lack of e-champions). Looking at experience in the United States, a report on the development of local e-governance by the Center for Technology in Government highlighted four key lessons based on experience with on-line public participation from 1993 to 1999 (Dawes et al. 1999). These can be summarized as follows:

- Information technology projects need to be driven by programmatic goals, not by technology. If the outcome is to improve service performance or ensure more effective delivery of information, then this should remain central, and potential management and policy implications should be fully evaluated.
- Government-supported information technology innovation for public participation should be approached from a learning perspective. Emphasis should be placed on the development of prototypes that can evolve, be evaluated and eventually grow.
- Government complexity needs to be addressed. Successful information technology projects require buy-in from different stakeholders within and outside of local government.
- Professionalism and personal commitment are essential for success in on-line public participation projects.

The report recommended that these lessons be addressed at the start of information technology projects to ensure a culture in government that encourages innovation, fosters experimentation and values thoughtful analysis.

The importance of considering local stakeholders in the development of local public participation projects based on information technology should not be underestimated. It is clear that local non-governmental organizations (NGOs) and communities face similar problems to those of the administration as they try to adapt to new demands related to the emergence of the Information Society. A 2001 study by the Surdna Foundation indicated that long-term structural changes induced by information technology are just over the horizon for the non-profit sector and that this process will change how they work, how they reach their audience, how they deliver on their goals and how they raise funds (Surdna Foundation 2001). Similar changes are taking place with on-line communities related to specific issues such as the environment. This initial experimentation is based very much on geographical locations and existing (rather than virtual) communities, although this might not remain the case for long. For instance, the 2001 Report on Online Communities by John Horrigan at the Pew Internet and American Life Project found:

People go online to connect to groups that have something to do with the place in which they live, with 29% of Internet users having at one time or another contacted a local community group or association and 30% having used the Internet for some involvement with a local charitable organization. (Horrigan 2001: 22)

A good example is the Seattle Community Network (SCN) established in 1995 by the local chapter of the Computer Professionals for Social Responsibility.[5] The SCN provides local environmental organizations with access to a number of on-line interactive tools, including telnet login, web mail, calendars, mailing lists, web hosting and volunteer opportunities support infrastructure. Another interesting example is the Minnesota E-Democracy group, a non-partisan citizen-based organization established in 1994,[6] whose mission is to improve participative democracy in Minnesota through the use of information networks (Browning 1996). It seeks to increase citizen participation in elections and promote public discourse on a range of issues through the use of the Internet.

Experience from these community-based and government-initiated activities suggests that the future of public participation is likely to be shaped by the forces promoting the digitization of governmental information, as well as service improvements, and by the traditionally countervailing civil society forces promoting participation and citizen empowerment. Significant progress has been made already with the development of the basic infrastructure, and interesting examples of e-environmental governance can already be found.

E-inclusion within the framework of e-governance

When we look into existing experience of e-inclusion in industrialized nations, particularly in relation to environmental governance, we do find a number of interesting similarities. We find, for instance, that greater accessibility to information has been accompanied by calls from many sectors for increased on-line interactivity and citizen participation. When looking at basic website usability, for example, a survey undertaken in 2000 by Darrell West at Brown University on e-governance at the state and federal level in the United States identified a number of important trends. The survey covered over 1,800 governmental websites and important findings include:
- only 5 per cent of government websites show some form of security policy and 7 per cent have a privacy policy;
- 15 per cent of government websites offer some form of disability access;
- 22 per cent of government websites offer at least one on-line service;
- 91 per cent of the sites responded to a sample email requesting the

official office hours of the particular agency, and three-quarters did so within one business day;
- in general, federal government websites did a better job of offering information and services to citizens than did state government websites.

The report concluded that "the e-government revolution has fallen short of its potential. Government websites are not making full use of available technology, and there are problems in terms of access and democratic outreach. E-government officials need to work to improve citizen access to online information and services" (West 2000: 1). Although we can assume that some progress has been made since 2000, the general impression is that we still have a long way to go before we will begin to see e-governance reaching higher quality levels. Another study (also published in 2000) on environmental democracy and environmental governance at the state level in the United States evaluated the performance of local government environment websites against a set of criteria related to access to information on the state of the environment and regulations, as well as interactivity in terms of citizen input, comment and communication via the website (Beierle and Cahill 2000). The report concluded that few of the 50 states surveyed have quality opportunities for interactive electronic public involvement. In some instances, local officials expressed serious reservations about the possibilities of increased interaction for the following reasons:
- on-line initiatives affect the internal organization of bureaucracies, requiring increased coordination and cooperation;
- responding to the external demands of stakeholders forces agencies to be strategic in their use of resources for on-line efforts;
- these demands for internal prioritization create tensions between departments and as a result engaging citizens on-line appears to be a considerably lower agency concern than streamlining the process aimed at the regulated community.

Similar studies are under way in Europe, including a major research project undertaken from 1998 onwards by the European branch of the International Council for Local Environmental Initiatives. The project, called ICTULA (Information and Communication Technology Use with Local Agenda 21), explores experience with the use of the Internet to support local environmental policy-making in five European cities – Amsterdam, Darmstadt, Hanover, Liverpool and Turku. Initial findings from an associated survey of 52 European local authorities found that 58 per cent were using the Internet to support their work, 21 per cent were using email to support Local Agenda 21 networking[7] and 33 per cent were using web pages to support Local Agenda 21 (ICTULA 1998). Looking specifically at experience in Darmstadt and Hanover, a number of risks were identified associated with the use of the Internet in terms of a resultant flood of information, possible alienation of interpersonal contacts, acceleration of all processes and the rise of new dependencies (e.g. if it can't be done without information technology, it won't be done). The benefits associated with Internet use in the context of Local Agenda 21 were highlighted as the potential for greater citizen involvement, opportunities for local authorities to share experiences rapidly, and new options for the coordination of local activists.

The overall impression from these studies is that e-governance is making only incremental inroads into improving public participation practices. The concern raised by some is that activities under the banner of e-inclusion, e-governance and e-democracy might actually be counterproductive. This thorny question is tackled by Riley (2004), who asks "is there really a 'democratic deficit' created (or maintained) from the way the digital divide works within the political system?" He goes on to respond that, "to the extent that the digital divide excludes those on the wrong side of it from good jobs and improving prospects, we have already shown that the answer is yes. However, many consider that argument to be indirect – in other words, the political consequences are the result of economic conditions rather than 'direct' political preference. Cases can always be found of low-income or digitally unconnected constituents who are very active politically, and of higher-income and digitally connected constituents who are not at all active politically and apparently have no desire to be so." He concludes by stating that "eDemocracy has not proliferated as widely or extensively into the public domain as many pundits had predicted just a few short years ago".

There is some empirical evidence on this topic, again from the United States and derived from the University of Southern California report entitled *Surveying the Digital Future – Year Four*, published in 2004. This report summarizes the findings from four annual surveys of households in the United States, around 2,000 each time. With respect to the political ramifications of Internet use, the surveys found that:

When asked, "by using the Internet people like you can better understand politics", more than half of Internet users (53 percent) in Year Four of the Digital Future Project agreed or strongly agreed – the highest level in the four years of the study (46 percent in 2002, 45.1 percent in 2001, and 46 percent in 2000).

When asked, "by using the Internet people like you can have more political power", the percentage of users who agree or strongly agree has fluctuated only modestly in all four years of the study: 27.3 percent in Year Four, 24.5 percent in 2002, 25.6 percent in 2001, and 30.0 percent in 2000. In the current study, 39.5 percent disagree or strongly disagree that the Internet can give people more political power.

When asked, "by using the Internet people like you will have more say about what the government does", 20.7 percent agreed or strongly agreed – about the same as the 19.9 percent of users in 2002 and 20.9 percent in 2001, and slightly less than the 24.2 percent in 2000.

Another survey in 2004, this time by the Pew Internet and American Life Project and covering 2,925 citizens, found that:

72% of Internet users contacted the government in the past year. This compares with 23% of non-Internet users in the past year.

Among Internet users, 30% say they have used email or the Internet to try to change a government policy or influence a politician's vote on a law. (Horrigan 2004: iii)

So the evidence is inconclusive but appears to suggest that a small and growing proportion of the population uses the Internet for political participation. This led Riley to conclude that perhaps "extended access would lead to somewhat more e-democracy, but expecting anything more than a small increase is exactly the kind of exaggerated prospect that both domestic and international experience shows to be unrealistic".

Concluding remarks

The review presented in this chapter clearly indicates that an increasing number of policy makers and researchers around the world are currently working valiantly to link information and decision-making with global trends and local needs. They are reflecting upon the pressing global problems facing modern communities and examining ways in which practical measures can contribute to understanding and amelioration of existing problems. The opportunities (virtual and real) associated with this new electronic interdependence truly reflect Marshall McLuhan's "Global Village" (McLuhan and Fiore 1968). To put it simply, globality implies the coming together of local cultures, a process that has become known as "glocalization". This is not an entirely neutral development and, as Zygmund Bauman (1998) so clearly explains, both globalization and localization can be understood as expressions of new polarizations and stratification in society. Nowhere is this more apparent than with respect to the emergence of the Information Society and the Internet.

It is clear that there could be many potential positive impacts of Internet use in support of public participation related to environmental issues. Moreover, there could be an additional bonus – when the Internet is used rapidly to internationalize examples of good practice through on-line networks and the creation of associated web-based epistemic communities. On the negative side, Internet use is likely to bring advantages only for the digitally connected, and many governments, already strapped for funds, will struggle to expand accessibility for their citizens. On-line public participation is not different from off-line versions. The same age-old problems have to be tackled, including how to develop trust and credibility. Moreover, there is the issue of how to reach those traditionally less active or the so-called "middle many", who could influence the process in a positive manner if they had the incentive to get involved. On top of this, we can anticipate the need, especially in the environmental arena, to explain complex information. As with all public participation, a clear communication strategy, responsive to local needs, is essential.

At the same time, there is a darker side to this new world of e-governance that we have not had time to delve into here. This includes the ever-growing concerns related to hacking, spamming, security, privacy, identify theft, propaganda and misrepresentation related to the use of information technology and the Internet (Browning 1996). We need to develop measures to deal with these as we progress in our application of information technology to support governance and public participation. In closing, we need to bear in mind, as Hills and Hughes found in their studies of US politics, that the Internet is not going to change governance or public participation radically. Rather, people are likely to mould it to their own ways of thinking and action – it is merely "a new venue for the same old human compunction: politics" (Hills and Hughes 1998: 186).

Notes

1. Thomas Jefferson to Richard Price, 1789. ME 7:253, accessible on-line at ⟨http://www.randycrow.com/articles/120800.htm⟩.
2. The key phrase here is "first step". The other higher steps include consultation, partnership, delegated power and citizen control in ascending order of importance.
3. For instance, a worldwide survey in 1996 by Reuters found that two-thirds of managers suffer from increased tension and one-third from ill health because of information overload (Reuters 1996).
4. See ⟨http://www.internetworldstats.com/stats.htm⟩.
5. See ⟨http://www.scn.org/⟩ (accessed 15 November 2003).
6. See ⟨http://www.e-democracy.org/discuss.html⟩ (accessed 15 November 2003).
7. Local Agenda 21 was first described in Chapter 28 of *Agenda 21*, the global action plan to promote environmental sustainability that was agreed at the 1992 United Nations Conference on Environment and Development (the Rio Earth Summit). This chapter called upon all local authorities to consult with their communities and develop and implement a local plan for sustainability – a "Local Agenda 21".

REFERENCES

Arnstein, S. (1969) "A Ladder of Citizen Participation", *Journal of the American Institute of Planners* 8(3): 217–224.
Barrett, B. F. D. and I. Yamada (2000) "Exploring the Global–Local Axis: Telecommunications and Environmental Sustainability in Japan", *Greener Management International*, Issue 21: 89–102.
Barrett, B. F. D., J. Fein, A. Kuroda and I. Yamada (2001) *E-Learning for a Sustainable World – Obfuscating the Real and the Virtual?* UNU/IAS Working Paper, Tokyo, May.
Bauman, Z. (1998) *Globalization – The Human Consequences*. London: Polity Press.
——— (2000) *Liquid Modernity*. Cambridge: Polity Press.
Beierle, T. and S. Cahill (2000) *Electronic Democracy and Environmental Governance: Survey of States*. Washington, DC: Resources for the Future.
Bowers, C. A. (2000) *Let Them Eat Data – How Computers Affect Education, Cultural Diversity and the Prospects of Ecological Sustainability*. Athens, GA: University of Georgia Press.
British Council (1999) *Developments in Electronic Governance, Information Services Management*. Manchester: The British Council.
Browning, G. (1996) *Electronic Democracy – Using the Internet to Influence American Politics*. Wilton, CT: Independent Publishers Group,
CEC [Commission of the European Communities] (2000) *E-Europe – An Information Society for All – Action Plan*. Prepared by the Council of the European Union and the Commission of European Communities for the Feira European Council, Brussels.
Dawes, S. S., P. A. Bloniarz, D. R. Connelly, K. L. Kelly and T. A. Pardo (1999) *Four Realities of IT Innovation in Government*. Albany, NY: Center for Technology in Government.
Devall, B. and G. Sessions (1986) *Deep Ecology*. Salt Lake City, UT: Gibbs Smith Publisher.
Fischer, F. (2000) *Citizens, Experts, and the Environment: The Politics of Local Knowledge*. Durham, NC: Duke University Press.
Heeks, R. (2001) *Understanding e-Governance for Development*. i-Government Working Paper No. 11. Manchester: Institute for Development Policy and Management.
Hill, K. A. and J. E. Hughes (1998) *Cyberpolitics – Citizen Activism in the Age of the Internet*. New York: Rowman & Littlefield.
Horrigan, J. B. (2001) *Online Communities: Networks That Nurture Long-Distance Relationships and Local Ties*. Washington, DC: Pew Internet and American Life Project.
——— (2004) *How Americans Get in Touch with Government*. Washington, DC: Pew Internet and American Life Project.
ICTULA (1998) *ICTULA: User-Expert-Dialogue in Darmstadt and Hannover*. Darmstadt: Institut für Zielgruppenmarketing und Kommunikation.
Katz, E., A. Light and D. Rothenberg, eds (2000) *Beneath the Surface: Critical Essays in the Philosophy of Deep Ecology*. Cambridge, MA: MIT Press.

McLuhan, M. and Q. Fiore (1968) *War and Peace in the Global Village*. New York: Bantam.
Mitchell, W. J. (2000) *E-Topia*. Cambridge, MA: MIT Press.
Negroponte, N. (1995) *Being Digital*. New York: Vintage Books.
Perri 6 (2004) *E-Governance: Styles of Political Judgement in the Information Age Polity*. London: Palgrave Macmillan.
Perri 6 with Jupp, B. (2001) *Divided by Information – The Digital Divide and the Implications of the New Meritocracy*. London: Demos.
Point Topic (2005) *World Broadband Statistics: Q4 2004*. London: Point Topic Ltd.
Reuters (1996) *Dying for Information? An Investigation into the Effects of Information Overload in the USA and Worldwide*, based on research conducted by Benchmark Research. London: Reuters Limited.
Riley, T. B. (2004) "E-Government – The Digital Divide and Information Sharing: Examining the Issues", paper prepared under the auspices of the Commonwealth Secretariat and co-sponsored by the Information Technology Resources Centre Public Works and Government Services, Canada.
Schenk, D. (1998) *Data Smog: Surviving the Information Glut*, revised and updated edition. San Francisco: Harper.
Shumacher, E. F. (1973) *Small Is Beautiful: Economics as if People Mattered*. New York: Harper & Row.
Slevin, J. (2000) *The Internet and Society*. Cambridge: Polity Press.
Surdna Foundation (2001) *More Than Bit Players: How Information Technology Will Change the Way Nonprofits and Foundations Work and Thrive in the Information Age*. Report by Andrew Blau to the Surdna Foundation, Inc.; available at ⟨http://www.surdna.org/documents/morefinal.pdf⟩ (accessed 15 November 2003).
Toregas, C. (2001) "The Politics of E-Gov: The Upcoming Struggle for Redefining Civic Engagement", *National Civic Review* 90(3): 235–240.
Turkle, S. (1995) *Life on the Screen – Identity in the Age of the Internet*. New York: Touchstone.
USC Annenberg School Center for the Digital Future (2004) *The Digital Future Report, Surveying the Digital Future – Year Four*. Los Angeles: University of South California.
West, D. M. (2000) *Assessing E-Government: The Internet, Democracy, and Service Delivery by State and Federal Governments*. Providence, RI: Brown University.

6

Promoting public participation in international waters management: An agenda for peer-to-peer learning

Dann M. Sklarew

All living beings, and humans in particular, need water to survive. As our water-dependent civilization grows in size and advances technologically, human water usage continues to intensify. The increasing threat of water usage conflicts has created a global mandate for governments to manage water for the common good.

Through the United Nations' Millennium Development Goals (MDGs), for instance, nations have agreed to improve their citizens' access to safe water for human health and sanitation, as well as to foster water's indirect benefits – through natural resource protection and agricultural and aquatic food security (United Nations 2000). These goals recognize water as fundamental to both human life and human rights. Improved management of both freshwater and marine resources is vital to ensuring these rights for a burgeoning world population in an equitable and environmentally sustainable manner.

In many cases, countries are rising nobly to this tremendous challenge: The post-apartheid South African constitution requires that the government ensure a minimum quantity of water for every one of its citizens (South Africa 1996). As part of its nationwide "no hunger" campaign, the current Brazilian administration also committed to realizing the goal of "no thirst" among its people (SRH 2003).

The MDG target for nations to establish Integrated Water Resources Management (IWRM) plans by 2005 further underscores national governments' leading role in ensuring their constituents' water benefits. Water management goals cannot be fully realized, however, without intimate understanding of local water needs, knowledge, usage and impacts for both mainstream and marginalized populations.

Those marginalized in governments' water decisions are frequently those most susceptible to its negative consequences. The result is a perpetuation of poverty, ecological vulnerability and increased potential for economic and political instability. People living in shantytowns along the Rio de la Plata in Buenos Aires, Argentina, for instance, frequently find their homes and shelters inundated, damaged without warning as floodwaters rise (Benavídez and Santoro 2004). Meanwhile, subsistence farmers and herders in rural Kenya vociferously protest against irrigation diversions, declining access to groundwater and ineffective resolution of their water scarcity concerns (Mkawale 2005; Mwangi 2005). Autocratic water policies, such as those regarding water rights, pollution impacts and fishing practices, have also increased destitution among disenfranchised populations around the world. Often these problems transcend national borders, increasing strife within and between countries sharing transboundary water systems.

Essentially all of the world's marine environments and 261 of the world's major river basins are shared by two or more nations. Since water knows no political boundaries, international waters (IW) management is vital wherever meaningful hydrological units (e.g. aquifers, river, lake and regional sea basins) transcend national boundaries (Sklarew et al. 2001). Consequently, numerous international agreements and initiatives have promoted such joint management of waters shared among nations (INBO n.d.; Wikipedia n.d.).

Numerous legal, institutional, cultural and political obstacles may constrain joint management of internationally shared water resources. For instance, national decision makers for internationally shared water resources are often spatially and politically far removed from the bodies of water they govern. When the United States signed its Boundary Waters Treaty (1909) with Canada and when the Democratic Republic of Congo signed the Lake Tanganyika Convention (2003), each country's capital was more than 1,000 km from their respective transboundary waters. Without sufficient local involvement, remote management of such international waters may result in international incidents. The transboundary river between Senegal and Mauritania illustrates this point: when government-supported dams changed the seasonal cycle of floods upon which poor farmers and migrating ranchers alike depended, violence erupted between the two groups and led to airlift repatriation of tens of thousands (personal observation). At the end of the twentieth century, border incidents of civil unrest also resulted from exclusive transboundary water development decisions from the Danube River in Europe to the Mekong River in South-East Asia.

How can nations cooperate to realize their respective water management agendas without alienating one another's or even their own constituents? This requires institutions capable of coordinating across international, national and community scales of governance within transboundary river basins and coastal areas.

There is a growing consensus among managers of international freshwater and marine systems that application of public participation (P2) and related stakeholder involvement is their top capacity-building priority (IW:LEARN 2002; Suarez and Sklarew 2002). A number of recent case studies in this volume and elsewhere have also demonstrated the importance of consciously proactive and inclusive P2 in managing international water resources (Bruch et al. 2005a; UPTW 2003; Chapter 1 in this volume). As Phillip Weller, executive director of the International Commission for the Protection of the Danube River (ICPDR), recently declared, it is important for international waters managers "to provide leadership in assisting countries and organizations in the region to strengthen the public understanding and commitment to a healthy [river] ... the pride and passion that people have for it needs to find expression" (ICPDR 2003). He further emphasized the opportunity for the basin organization to assist countries in developing processes that encourage critical dialogue among all stakeholders for achieving "equitable and economically and ecologically sound" basin management.

At a very basic level, members of the public have a fundamental right to be involved in decisions that have the potential to seriously affect their health, prosperity and well-being. Public participation seeks to ensure that citizens have the opportunity to be notified, express their views and influence these decisions.

The public are also a valuable source of information related to local natural resources and thus may provide critical insights into local ecosystem functions and health upon which their communities depend. Actively involving the public in decision-making also clarifies the values and trade-offs that communities associate with various uses of and impacts on their water resources. Moreover, including public participation from the outset of a programme or project often defuses potential opposition by providing a forum for developing solutions that are acceptable to many parties. Ultimately, public participation strengthens the democratic character of decision-making processes and helps build broad-based consensus on water resource use and management.

How should international waters managers pursue such opportunities?

Recognizing that each situation is unique, I and others assert that there nonetheless exists an emerging set of adaptable approaches and transferable techniques for increasing participation in international waters decisions (Bruch et al. 2005b; Chapter 9 in this volume). If well organized and shared across the international waters community, these participatory processes and mechanisms could help ensure more effective, equitable and enduring water management results.

This chapter presents an agenda and framework for peer-to-peer learning to promote effective public participation in international waters management. Building upon the previously noted studies and relevant public participation experiences from other domains, the proposed agenda includes a series of workshops developed within and for different geographical and cultural regions of the world. These peer-to-peer workshops will iteratively vet, derive and incorporate lessons from the varied IW initiatives in each region. This is a collaborative process, involving both governments and civil society. Together, successive regional cohorts of participants will develop an overall framework for evaluating, applying and strengthening public participation throughout an adaptive IW management process.

As a starting point, the proposed learning agenda should address the following common concerns:
1. What is public participation within the context of international waters management?
2. Why should international waters management incorporate public participation?
3. Who should foster these public participation activities?
4. At what point(s) should public participation be applied within an adaptive management process for international waters?
5. How can public participation be applied or adapted across diverse national and regional circumstances?
6. Which tools and techniques are appropriate for addressing common challenges to participatory international waters management?
7. Where can one go for additional assistance in such efforts?

A brief exploration of these issues below aims to provide a preliminary background for IW managers to consider in developing and implementing stakeholder involvement plans (SIP) to augment public participation within their transboundary basins. It furthermore serves as an embryonic framework to organize potential insights and lessons provided by participants in the proposed workshop series.

Following this discussion, I offer a more detailed description of the workshops and extend an invitation to readers to participate in the workshop development process. Thus, the seeds sown here may germinate into a quite distinct and, I hope, valuable product through your collective contributions.

What is public participation within the context of international water management?

Principle 10 of the 1992 Rio Declaration on Environment and Development first articulated the three defining "pillars" of public participation in environmental decision-making: (1) access to information; (2) access to participation in decision-making; and (3) access to justice in environmental matters; in other words:
(1) informing people of water management issues or activities that may affect them while ensuring transparency in the overall decision-making process;
(2) involving the public in decision-making regarding such activities; and
(3) providing those adversely impacted by these decisions and activities with means for seeking redress (United Nations 1992).
This functional definition has been applied by a variety of national, regional and global environmental partnerships.[1] Significant agreements at the international scale include the legally binding Aarhus Convention (UNECE 1998) and the recently revised African Convention on the Conservation of Nature and Natural Resources (African Union 2003).[2]

The seeds of participatory governance were planted long before these agreements. Thus evolved distinct regional perspectives. Still, across vastly different political and cultural contexts, public participation has been described in a variety of similar ways.

Guidance from the European Union's Water Framework Directive (WFD) defines public participation in bureaucratic terms as "allowing people to influence the outcome of plans and working processes" (European Union 2002). The Organization of American States (OAS) more broadly declares that public participation is "all interaction between government and civil society, and includes the process by which government and civil society open dialogue, establish partnerships, share information, and otherwise interact to design, implement, and evaluate development policies, projects, and programs" (OAS 2000). Thus, public participation includes government mechanisms to involve "people" and "civil society".[3] In Europe, these mechanisms are now enforceable by international law (UNECE 1998), whereas the Americas have agreed only to recommendations and policy guidance to encourage national-level implementation (OAS 2000).

As regards public participation [*gong zhong can yu*] in the world's most populous nation, *China's Agenda 21* White Paper views public participation as a responsibility that extends beyond decision-making alone: "It is necessary for the public to not only participate in policy-making related to environment and development, particularly in areas which may bear direct impact on their living and working communities, but also to supervise the implementation of the policies" (China SEPA 2001).[4]

The Arab League views participation as a means to pursue societal progress. Its Tunis Declaration asserts that members should endeavour to reform, modernize and adapt to a rapidly changing world by "enlarging participation in public and political life" (Arab League 2004). This includes promoting a P2-enabling environment through respect for human rights, freedom of expression and judicial independence, as well as "fostering the role of all components of the civil society, including NGOs" and "widening women's participation". Legal mechanisms for enabling public participation have only just begun.

Meanwhile, the African Charter for Public Participation views P2 both instrumentally and more profoundly as being a "fundamental right of the people to fully and effectively participate in the determination of the decisions which affect their lives at all levels and at all times" (UNECA 1990). This value is also recognized in the revised African Convention, which includes public participation as an essential procedural right to be included in ratifying nations' domestic legislation (African Union 2003).

Public participation proponents often reach out to marginalized members of the public by explicitly including civil society, the poor, ethnic and religious minorities, local and indigenous communities, women and children – collectively a majority of all humanity. This public is usually distinguished from stakeholders in general, which include more powerful institutions, such as political, economic and religious élites; international donors; and government agencies beyond the primary decision-making authority itself. Thus, a related term, "stakeholder involvement", may be used to consider involvement both of those affected by a decision (the public) as well as of those in positions to influence its intended outcome (the powerful).

Following the Rio Declaration, international donors and governments have increasingly expected "top–down" promotion of "bottom–up" participation in international waters management and related efforts. The Global Environment Facility (GEF), a major global catalyst for IW financing, requires that all of its projects "provide for full disclosure of non-confidential information, and consultation with, and participation as appropriate of, major groups and local communities throughout the project cycle" (GEF 1996). Thus, public participation foci for GEF projects include information dissemination, consultation and "stakeholder

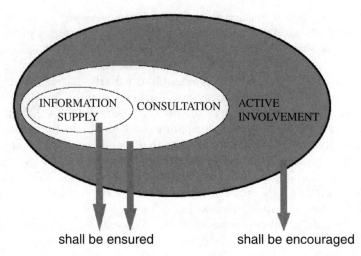

Figure 6.1 Public participation guidance from the EU Water Framework Directive.
Source: European Union (2002).

participation" (i.e. collaboration throughout the project cycle). With respect to Europeans' water policy-making, Figure 6.1 illustrates that their WFD public participation guidance also emphasizes that disclosure and consultation are essential and active involvement is to be encouraged (European Union 2002).

The World Bank (IBRD) targets a more detailed a set of operative P2 approaches across a spectrum of public influence and impact: (a) disclosure/information-sharing (one-way communication); (b) dialogue/consultation (two-way communication); (c) collaboration (shared control over decisions and resources); and (d) empowerment (transfer of control over decisions and resources) (Avramoski 2004; World Bank 2004). The Bank also requires stakeholder consultation in all environmental assessments for projects it supports, and disclosure where adverse impacts are likely (World Bank 2000). Other institutions have developed similar spectra of participation approaches, where each level is associated with distinct goals, commitments and tools to enhance participatory decision-making towards an effective outcome (e.g. UNDP 1997; IAP2 n.d.(b)).[5] As a result, any or all of these approaches may be appropriate within specific policy-making contexts (e.g. issues of concern and their underlying causes, urgency of decision, existing legal frameworks for P2, awareness, and commitment of potential participants).

Because 40 per cent of the Earth's population live within basins that cross national borders, addressing water challenges is frequently a trans-

boundary effort. In the context of the Millennium Development Goals for sustainable development and poverty alleviation, public participation is applicable to a variety of international waters challenges. Among these are, for instance:
- freshwater scarcity
- freshwater and marine pollution
- habitat and community modification
- unsustainable exploitation of fisheries and other aquatic living resources
- global change (e.g. sea-level rise, ozone depletion, water and nutrient cycling) (GIWA n.d.).

Behind problems associated with each of these challenges exists a causal chain of activities, a cascade of linkages back to their underlying root causes. Addressing such concerns requires an iterative process of *adaptive management*, including characterization and prioritization of IW problems, investigation of their root causes, identification of viable options for abatement, selection and implementation of preferred actions, and assessment of progress towards ecological and social improvement.

Under appropriate conditions, public participation can play an important role in each and every stage of adaptive management as applied to international waters.

Why should international waters management incorporate public participation?

Everyone has an interest in access to water, yet many are still without such access

We are water beings living on a water planet. We each consume 2–4 litres of water per day, accounting for 60 per cent of our body weight (Howard and Bartram 2003). We are thus utterly dependent upon a sufficient, regular supply of drinkable water – i.e. without excessive salts, pathogens or natural or artificial toxins. Optimal daily water access for all human consumption and hygiene each year would be equivalent to 14–28 per cent of the Earth's total river volume and less than 0.5 per cent of its volume of either lakes or aquifers.[6] Beyond such usage, water is also essential to food production (agriculture and fisheries), electrical power, industry, navigation and recreation – altogether using a tiny fraction of the surface water available to support most life and the ecological processes that sustain us.

Yet water is unevenly distributed over our planet's surface and highly variable over time. Even where there is enough, distribution to those in need is catastrophically lacking. Over 1 billion people today lack sufficient access to safe water resources (WHO 2003). Water for subsistence is a fundamental input for their poverty alleviation. Moreover, overfishing, rapid coastal development and the accompanying increases in pollution levels in the world's marine ecosystems are increasingly undermining the sustainability of coastal communities. These tragedies extend across parts of the Americas, Africa and Asia to the south-west Pacific islands (Gleick 1998). In some places, we also live too close to water – such as in the floodplains of Bangladesh or coastal areas along the Caribbean islands – where millions are regularly vulnerable to flooding. As the terrible Indian Ocean tsunami demonstrated in December 2004, even populations that now appear secure one day may become vulnerable again the next. Each and every one of us has a keen interest, or *stake*, in living within range of reliable access to water without drowning in its deluge.

Decisions affecting international water – how to allocate water supply among riparian countries or how to manage the impacts of developing shared fisheries – only rarely reflect the interests of the populations living in national border areas. As noted above, these populations are frequently far from the sources of power. In order to be able to advocate effectively for their water-related needs and priorities, their communities, local governments and civil society organizations that represent them must have the tools and the capacity to participate meaningfully in the decisions that will determine the ways in which international water resources are allocated and managed.

With population growth and no public participation, some people's use of water resources is increasingly likely to limit or negatively affect others

Exponential growth in human population, greater settlement sizes and density and technological advancement have increased the aggregate ecological risks associated with human activity. This so-called "Tragedy of the Commons" (Hardin 1968) has particular relevance to water resources, given the degree to which surface waters are recycled and reused throughout their long-term path from air to sea. Thus, population and associated agricultural growth, in places such as the south-western United States or eastern China, have resulted in large public works projects to redirect rivers over vast distances, create massive irrigation networks, and store and release tremendous quantities of water over lands previously inhabited by millions.

To view the potential impacts when such efforts go awry, one need look no further than the Aral Sea catastrophe (Kriner 2002). The Aral Sea was historically one of the Earth's great inland waters, but it has since been reduced by two-thirds owing to river diversion for increasingly intensive farming over the past century (LakeNet n.d.). Today, the sea has been split into two much smaller lakes within a wasteland of contaminated sediments. One wonders if such a tragedy might have been averted if the fishing and riparian communities had been adequately included in the Soviets' agricultural decision-making.

Without conscientious attention to the knowledge and interests of those most affected by such activities, population growth and advancing technologies for mobilizing natural resources could result in much larger, more profound impacts on society and its life support systems. This holds for freshwater shortages, as well as for over-fishing and the additional IW challenges identified above.

National governments cannot do it alone

Effective management of water resources requires concerted efforts at personal, community, provincial, national and international scales, across nearly every sector of society. As a result, as Parr (2005: 26) notes with respect to Thai natural resources management, "government operating alone 'fails to reach the goal of efficient and sustainable rehabilitation, restoration and exploitation'". Government limitation is also illustrated by the Senegal River, Aral Sea and other cases presented here. Thus, participatory processes are critical to governments' realizing their own missions for environmental management in general, and water management in particular.

Water cycles and flows across political boundaries, as do water-related concerns, thus necessitating a transboundary approach to public participation

As water circulates above, on and below ground, negative impacts from water and related natural resource usage may extend far from the source or root cause of the problem. Pollution-induced acid rain problems in North America, Europe and north-east Asia and impacts of withdrawals from shared aquifers in Africa and South America, for instance, typically extend well beyond national boundaries. Similarly, marine pollution frequently occurs on the high seas between continents. Thus, the affected public may include populations of stakeholders far removed from the root cause of IW concerns, their interests pertinent to any distant IW management decision that affects them.

We are also indirect, long-distance beneficiaries with respect to the Earth's international waters. As one example, humanity consumed over 60 million tons of fish in 2001, over one-third of which was delivered across international borders (Vannuccini 2003). Furthermore, an equivalent of over 5 per cent of global fish "production" that year went from low-income food-deficit countries to developed countries, notably the European Union, Japan and the United States (ITC n.d.). Since fisheries depletion is a worldwide problem now (WSSD 2002), nations recognize the need to create transparent, participatory institutions to jointly engage neighbouring nations, fishing interests and non-governmental organizations (NGOs) in sustainable fisheries management (e.g. WCPFC 2000). Realizing UN goals to establish a network of marine sanctuaries to help restore fisheries worldwide (WSSD 2002) will require similarly coordinated efforts to promote international cooperation among these varied types of stakeholders. International cooperation among various sectors of society is needed to restore, sustain and enlarge healthy coral reefs for fisheries protection as well as for added human benefits in terms of mitigating the coastal damage from typhoons and tsunamis (Benson 2005; Parr 2005).

Given the myriad transboundary interrelations alluded to above, IW management will succeed in addressing the root causes of degradation only by involving stakeholders across jurisdictions and economic sectors. This includes all relevant administrative and ecological units. Such an effort implies an increase in the complexity and coordination of public participation across national boundaries, transcending languages, economies and political systems – a significant capacity challenge for many IW management teams. Fortunately, recent experience from Europe (e.g. WFD, ICPDR), North America (IJC n.d.), GEF projects and elsewhere indicates that such challenges may be surmountable.

Who should foster these public participation activities?

In some cases, the public foster their own participation: P2 efforts may be instigated in reaction to public pressure or concern over past or pending government decisions with perceived adverse affects on local communities. Often this is the case when a government does not have in place the appropriate legal and institutional frameworks for promoting public participation prior to such confrontational manifestations. In such instances, parties recognized as unbiased by the government and by the affected communities – such as non-advocacy national or regional NGOs – may play a role in facilitating public participation activities.

Increasingly, however, governments themselves have recognized that public participation is critical to their continued good governance. As one Chinese newspaper advises, "Listening to the public's voice rather than adopting an ivory-tower approach is a wise way for government departments to win public understanding and avoid making problematic policies" (*China Daily* 2004). In such circumstances, governments at national, provincial and even local scales may be called upon to serve as institutional facilitators and underwriters of the process. The skills and resources are not always present, however (e.g. HELCOM 2003).

The Access Initiative (TAI) assessments indicate that implementing national public participation frameworks for environmental management is a multifaceted and challenging endeavour for any government (TAI n.d.). Thus, where skills and resources to support public participation are not readily available, international donors may be called upon to assist both financially as well as in terms of capacity-building. Civil society, both national and international NGOs, also has a great deal of experience and capacity to offer in this regard.

In other cases, as previously noted for the GEF, UNDP and World Bank, donor agencies expect "bottom–up" public participation to be embedded in their projects and offer guidance for doing so. The planned workshop series described in this chapter is a prime example of such guidance.

With respect to IW management, donors have an added responsibility to ensure that consultation and related P2 approaches extend beyond relatively affluent and/or democratic riparian states to include citizens of those countries that are less empowered. In this regard, international basin organizations have discovered that regional NGOs may play a vital role in outreach and involvement across their transboundary basin, especially where national governments lack either the capacity or the political will to engage stakeholders effectively (Chapter 9 in this volume).

Finally, civil society organizations of all types – from labour cooperatives to women's organizations, trade associations to religious groups, local universities to international NGOs – have within their mission to promote their constituents' well-being. Thus, each has a unique and sometimes unforeseeable role to play in promoting public participation to enhance international waters management.

At what point(s) should public participation be applied within an adaptive management process for international waters?

The EU Strategic Environmental Assessment Directive suggests that information-sharing, consultation and involvement should occur throughout any adaptive management regime (European Union 2003). Duda and Uitto (Chapter 9 in this volume), meanwhile, highlight how the adaptive management process lends itself to these sorts of public participation activities. In particular, they describe the Global Environment Facility's approach of using transboundary diagnostic analysis (TDA), embedded within the process of developing and implementing a strategic action programme (SAP), as the initial iteration of an adaptive management regime.

The TDA/SAP process has been preliminarily summarized by Bloxham

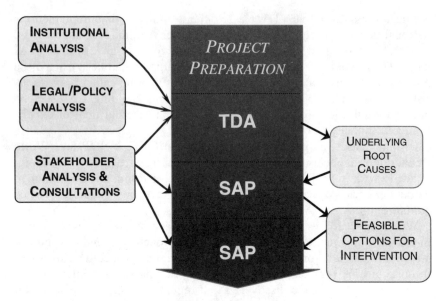

Figure 6.2 Key stages in a TDA/SAP process.
Source: Adapted from Bloxham and Mee (2004).

and Mee (2004), as adapted in Figure 6.2. Stages include: (1) developing a project concept, (2) joint fact-finding, (3) preparing the SAP, and (4) implementing the SAP. From the initial stages of project conception, the facilitator is advised to identify and consult with stakeholder groups. IW management may continue to benefit from increasing integration and institutionalization of public participation from conception through SAP implementation. In the TDA development phase, for instance, stakeholder consultation is expected to include: (a) stakeholder analysis through interviews and questionnaires, which also informs concurrent institutional and legal/policy analyses; (b) development of a public involvement plan; and (c) stakeholder review of the draft TDA (Bloxham, personal communication). As Duda and Uitto note in Chapter 9, this could involve providing local knowledge to support joint fact-finding activities, partnering in demonstration projects and assuring accountability though citizen-based monitoring and evaluation exercises. Encouraging civil society to develop a daisy-chain of public outreach and awareness-raising has often been effective across many basins.

Further investigation in preparation for the workshop series should clarify the specific stages and means by which GEF IW projects and similar IW management institutions have incorporated public participation into the SAP development process and parallel initiatives elsewhere.

How can public participation be applied or adapted across diverse national and regional international waters circumstances?

The GEF (1996) asserts that "effective public involvement should enhance the social, environmental, and financial sustainability of projects". Its public involvement guidance also provides a larger set of values to include in its environmental projects. Other principles have been summarized by the UNECE-based Conventions for Transboundary Waters (Helsinki Convention, UNECE 1992) and Public Participation (Aarhus Convention, UNECE 1998) – both of which are open to ratification by nations outside of Europe – as well as in the European Union's Water Framework Directive.

Public involvement initiatives should be based on a set of culturally and politically relevant principles such as these to ensure that the public participation means justify the IW management ends. The International Association for Public Participation (IAP2) provides a series of valuable training sessions to guide practitioners within affluent, democratic societies. The above European sources, along with the African Charter (UNECA 1990), the revised African Convention (African Union 2003) and the Inter-American Strategy for public participation (OAS 2000), provide additional direction regarding how to adapt public participation to various regional and policy-making contexts. However, at present, there is little on-the-ground experience with adapting these regional approaches and evaluating their applicability across continents – a key facet of the workshop series proposed here.

Within specific transboundary basins, the TDA/SAP process also suggests an overall approach that includes identifying stakeholders and involving them in basin-wide awareness-raising and fact-finding, followed by participatory establishment of goals, measures and targets. Through informed consultation at basin-wide, national, provincial and local scales, such P2 can establish a public mandate for scaling up activities from donor-supported projects to indigenously sustained international waters management programmes.

Which tools and techniques are appropriate for addressing common challenges to participatory international waters management?

Many IW projects actively share information to pursue better IW decisions, plans and actions. Many tools exist to promote stakeholders' access to IW information, thereby increasing their understanding and helping to enable their participation in decisions that affect them. Awareness-raising and transparency in decision-making are two broad tools that can help build public capacity to participate meaningfully and usefully in IW decisions.

Public service announcements (advertisements), photo contests and video documentaries are just a few techniques used by GEF IW projects. Social marketing to affect attitudes and behaviour via traditional mass media – radio, television and periodicals – remains important to communicate with populations beyond the reach of the newer, yet less pervasive, Internet. FarmRadio, for instance, sends audio scripts on improving water conservation to local radio stations to adapt and read as local public service announcements. The South Pacific island-nation of Kiribati, meanwhile, is developing an entire advertising campaign for social marketing to meet its regional International Waters Project aims.

Also useful are educational materials and exhibits. For example, the "Black Sea Shell Palace" is a portable and inflatable exhibit, with puppetry and other recreational learning elements. Basin-wide events are also valuable in instilling appreciation of shared waters – "Danube Day", for instance, when communities enjoy river-focused activities across many locations on one day each year. Such outreach increases public awareness and contributes to public demand for access to meaningful IW information. More in-depth information access tools include "help desks" accessible via mail, phone, fax or email; annual "state of the basin" reports in straightforward language to interest non-technical readers; as well as posters, maps, graphs and databases of basin-wide pollution inventories, degradation "hot spots" and "sensitive areas". For more interactive and extensive information-sharing, IW managers use tools such as citizen advisory groups, NGO forums, information centres, associations of public water users, and roundtables between industry and the public.

Participation in decision-making often begins with a conscientiously developed stakeholder analysis and stakeholder involvement plan, further refined with input from representatives of local public interests. The process also includes a vital stakeholder analysis of public needs and perspectives. In fact, such assessment may challenge decision makers' assumptions regarding priority environmental concerns, as recently occurred with the Caspian Sea Environment Programme. As stakeholder needs are identified and understood, and in order to continue public consultation and involvement in resource decision-making indefinitely, specific institutional and programmatic mechanisms develop, such as local water advisory boards, citizen review committees, and transboundary public preview and comment periods prior to implementing policies or permits that alter transboundary water quality, quantity or habitat.

Environmentally, economically and socially oriented NGOs are often eager to contribute to implementing IW management decisions. To foster such collaboration, IW managers may help convene NGO forums, underwrite the costs for developing shared visions, or invite NGOs to apply for small grants for demonstrating local restoration, education or awareness-raising. Inclusive NGO partnerships may even be cost-efficient for the formulation, delivery and replication of IW demonstrations.

To promote environmental justice across international boundaries, the Helsinki Water Convention mandates that citizens of affected nations have the same rights to redress as do the citizens of the nation where the impact was generated (UNECE 1992). Some transboundary basin organizations, such as the one between Finland and the Russian Federation, have also established legal and institutional mechanisms to address transboundary concerns raised by their constituents. Overall, however, institutional support for transboundary access to justice in water resources management is still in its infancy.

Where can one go for additional assistance in such efforts?

A plethora of public participation guides and tools is available from the references cited in this chapter (e.g. IAP2, WFD P2 guidance, UNDP, World Bank, ICPDR). In addition, the Aarhus Convention has a growing on-line library of documents related to P2 for environmental management.[7] The IAP2 (on-line) and World Bank participation materials[8] are a few of other many on-line sources to help learners to build capacity in public participation. However, few focus specifically on P2 for IW management, and none that do so are on a multi-regional or global scale.

With respect to public participation for IW management, in particular, the International Waters Learning Exchange and Resource Network (IW:LEARN) has established a collaborative clearinghouse for the IW community, which is accessible on-line at: ⟨http://participation.iwlearn. net⟩. This site expected to grow as additional materials are identified, assembled and produced during the course of the workshop series' development and delivery.

Realizing the learning agenda for public participation in international waters management

Since 1996, the GEF has been dedicated to "facilitate the exchange of best practices on public involvement" (GEF 1996). It has recently provided financial support and direction to IW:LEARN to develop the public participation for IW management workshop series over the 2004–2008 period. To do so, IW:LEARN is working in close cooperation with the Environmental Law Institute (ELI), the UNECE Helsinki and Aarhus

Convention secretariats and the Lake Peipsi Centre for Transboundary Cooperation (CTC), as well as other global and regional partners.

The series was introduced through an orientation session for IW managers at the 3rd GEF International Waters Conference in Salvador, Brazil (June 2005). That same month, the first regional workshop was held in St Petersburg, Russian Federation. At that workshop, over 65 participants from Eastern Europe, the Caucasus and Central Asia conveyed the following:
- officials make better decisions when the public are included;
- public participation includes both public outreach (awareness-raising) and public input (consultation);
- the public is multifaceted and broad;
- NGOs can valuably contribute to the water management solution;
- the participation process is gradual, messy and challenging;
- tools exist to improve participation across institutions, programmes and events;
- success is far off but progress is real.

After IW:LEARN further assesses and documents the lessons from St Petersburg, the series will proceed iteratively through Latin America, Africa and South-East Asia. Through each iteration, additional lessons will be derived and practices documented to create global guidance for applying P2 to IW management.

The workshop coordination team is conducting in-depth research on process and relevant legal, policy and management tools that can be used to increase effective public participation in IW management. The resources referenced in this article, other case studies and participants' own experiences are all contributing to this effort. This is expected to result in the following:
- materials to introduce the initial public participation curriculum, vetted by regional needs assessment, incorporating feedback into the materials before delivering subsequent versions of the modules;
- open content, collaborative documents on P2 for IW management, disseminated widely across various capacity-building networks (Cap-Net affiliates, UPTW, UNESCO water centres, INBO, LakeNet, International Shared Aquifer Resource Management);
- strategic planning for perpetual enhancement and delivery of public participation for IW management training as needed worldwide.

In this way, the workshop development process itself demonstrates both the value of diverse perspectives and the power of broad collaboration and outreach.

To reach a broad audience beyond the classroom, approaches and tools characterized throughout the series are also being incorporated into a guidance document or "toolkit" that distils the lessons from the initial research phase, elaborates a diverse cross-section of ground-tested and innovative participatory approaches and mechanisms, and highlights global similarities as well as regional or national distinctions. Partners will disseminate the guidance document at various points via the Internet, CD-ROM and hardcopy versions in the principal United Nations languages. Electronic formats will be used to encourage participants and other practitioners to adapt and translate training materials to their local languages and specific international waters circumstances. Thus, in addition to advancing stakeholder involvement plans and actions in participating basins, the initiative will also produce a living record for applying public participation to improve international waters management.

IW:LEARN and its partners invite IW managers, interested members of the public and private sectors, and civil society at large to participate in the workshop series design, development and evaluation. To do so, please contact me as follows: Dann M. Sklarew, Director and Chief Technical Advisor, IW:LEARN, 1638 Connecticut Ave, NW, Suite 310, Washington, DC 20009, USA; email: dann@iwlearn.org; tel: +1 (202) 465 4600; fax: +1 (702) 552 6583.

Acknowledgements

I would like to thank Jessica Troell of the Environmental Law Institute and Janot-Reine Mendler de Suarez of IW:LEARN for their insightful review of this chapter and their recommendations.

Notes

1. An international association of NGOs recently formed The Access Initiative (TAI) to monitor national efforts to promote Principle 10-driven public participation in environmental decision-making (TAI n.d.).
2. A general discussion of public participation in environmental management is provided in other chapters of this volume (e.g. Chapter 1) and elsewhere, and is beyond the scope of this article.
3. The North America-based International Association for Public Participation elaborates by equating public participation with "any process that involves the public in problem solving or decision making and uses public input to make better decisions" (IAP2 n.d.(a)).
4. The Aarhus Convention and related European agreements also promote participation at other points in the environmental management process, i.e. beyond the decision-making moment itself.
5. IAP2 further distinguishes between "consultation" (obtaining feedback), "involvement" (considering public issues and concerns) and "collaboration" (true partnership).
6. Calculated using the product of an annualized estimate from 100–200 litres/day optimal access (Howard and Bartram 2003) and a world population of 6.4 billion circa January

2005 (United States Census Bureau 2005), then divided by total water volumes found in rivers (1,700 km^3), lakes (100,000 km^3) and groundwater (8.2 million km^3) as provided in "Earth's Freshwaters" (1996).
7. See ⟨http://aarhusclearinghouse.unece.org⟩.
8. See ⟨http://www.worldbank.org/participation/⟩.

REFERENCES

African Union (2003) *African Convention on the Conservation of Nature and Natural Resources (Revised Version)*, ⟨http://www.africa-union.org/home/Welcome.htm⟩.

Arab League (2004) *Tunis Declaration*. 16th session of the Arab Summit, Tunis, Tunisia, 22–23 May.

Avramoski, O. (2004) *The Role of Public Participation and Citizen Involvement in Lake Basin Management*. Lake Basin Management Initiative Thematic Paper. Annapolis, MD: LakeNet; available at ⟨http://www.worldlakes.org/programs.asp?programid=2⟩.

Benavídez, F. and A. Santoro, directors (2004) *El Zanjón ("The Big Ditch"): A Short Film about Climate Change, Adaptation and Development in an Argentinean Shantytown*. Produced by Pablo Suarez (suarez@bu.edu).

Benson, J. (2005) "Professor Puts Knowledge to Good Use in Tsunami-Ravaged Thailand: Pomeroy Asked to Help with Recovery of Coastal and Fishing Communities", *The Day* (New London, CT), 19 January.

Bloxham, M. and L. Mee (2004) *The Use of GEF Processes for Collaboration on Transboundary Waters. Commission on Sustainable Development, 12th Session (CSD-12)*. New York: Learning Centre, 30 April.

Boundary Waters Treaty (1909) *Treaty between the United States and Great Britain Relating to Boundary Waters, and Questions Arising between the United States and Canada*, ⟨http://www.ijc.org/rel/agree/water.html⟩.

Bruch, C., L. Jansky, M. Nakayama, K. A. Salewicz and A. Z. Cassar (2005a) "From Theory to Practice: An Overview of Approaches to Involving the Public in International Watershed Management", in C. Bruch, L. Jansky, M. Nakayama and K. A. Salewicz (eds), *Public Participation in the Governance of International Freshwater Resources*. Tokyo: United Nations University Press.

Bruch, C., L. Jansky, M. Nakayama and K. A. Salewicz, eds (2005b) *Public Participation in the Governance of International Freshwater Resources*. Tokyo: United Nations University Press.

China Daily (2004) "Public Input Crucial in Formulating Policies", *China Daily*, 5 August; available at http://www.china.org.cn/english/government/103072.htm.

China SEPA [State Environmental Protection Agency] (2001) *China's Agenda 21 – White Paper on China's Population, Environment, and Development in the 21st Century*, ⟨http://www.zhb.gov.cn/english/SD/21cn/write_paper/index.htm⟩.

"Earth's Freshwaters" (1996) in *Macmillan Encyclopedia of Earth Sciences*. New York: Simon & Schuster; cited at ⟨http://hypertextbook.com/facts/2000/VanessaBallenas.shtml⟩.

European Union (2002) *Water Framework Directive Guidance Document on Public Participation*, ⟨http://forum.europa.eu.int/Public/irc/env/wfd/library?1=/framework_directive/guidance_documents/participation_guidance&vm=detailed&sb=Title⟩.

────── (2003) Directive 2001/42/EC of the European Parliament and Council of June 27, 2001, on the assessment of effects of certain plans and programmes on the environment. *Official Journal of the European Communities*, L197/30 [EN]; available at ⟨http://europa.eu.int/eur-lex/pri/en/oj/dat/2001/l_197/l_19720010721en00300037.pdf⟩.

GEF [Global Environment Facility] (1996) *Public Involvement in GEF-Financed Projects*. Washington, DC: GEF; available at ⟨http://www.thegef.org/Operational_Policies/Public_Involvement/public_involvement.html⟩.

GIWA [Global International Waters Assessment] (n.d.) *Water Issues: Causal Chain Analysis*, ⟨http://www.giwa.net/caus_iss/causual_chain_analyses.phtml⟩.

Gleick, P. (1998) *The World's Water: The Biennial Report on Freshwater Resources*. Washington, DC: Island Press; available at ⟨http://www.worldwater.org/⟩.

Hardin, G. (1968) "The Tragedy of the Commons", *Science* 162: 1243–1248; available at ⟨http://dieoff.org/page95.htm⟩.

HELCOM [Baltic Marine Environment Protection Commission] (2003) Press Release: International workshop "Sustainable River Basin Management and Public Participation" in Tartu, Estonia, June 6; available at ⟨http://www.helcom.fi/helcom/news/302.html⟩.

Howard, G. and J. Bartram (2003) *Domestic Water Quantity, Service Level and Health*. Geneva: World Health Organization.

IAP2 [International Association for Public Participation] (n.d.(a)) *IAP2 Code of Ethics for Public Participation Practitioners*, ⟨http://iap2.0rg/boardlink/code-of-ethics.shtml⟩.

────── (n.d.(b)) *IAP2 Public Participation Spectrum* [chart], ⟨http://iap2.0rg/practitionertools/index.shtml⟩.

ICPDR [International Commission for the Protection of the Danube River] (2003) "Phillip Weller – the new Executive Secretary of ICPDR", *Danube Watch Magazine* 2003:2; available at ⟨http://www.icpdr.org/pls/danubis/docs/folder/home/icpdr/icpdr_doc_centre/danubewatchmagazines/dw2003_2/index.htm⟩.

IJC [International Joint Commission] (n.d.) *International Joint Commission – Commission Mixte Internationale*, ⟨http://www.ijc.org/⟩.

INBO [International Network of Basin Organisations] (n.d.) ⟨http://www.riob.org/friobang.htm⟩.

ITC [International Trade Centre] (n.d.) *Imports 1998–2002 – International Trade Statistics by Product Group*. UNCTAD/WTO, ⟨http://www.intracen.org/tradstat/welcome.htm⟩.

IW:LEARN [International Waters Learning Exchange and Resource Network], ed. (2002) *Summary Report of the Second Biennial Global Environment Facility (GEF) International Waters Conference, Dalian, China, September 25–29, 2002*, ⟨http://www.iwlearn.net/ftp/IWC2002_Final_Draft_Report.doc⟩.

Kriner, S. (2002) "Aral Sea Ecological Disaster Causes Humanitarian Crisis", *Redcross.org*, 10 April, ⟨http://www.redcross.org/news/in/asia/020410aral.html⟩.

LakeNet (n.d.) *Lake Profile: Aral Sea*, ⟨http://www.worldlakes.org/lakedetails. asp?lakeid=9219⟩.
Lake Tanganyika Convention (2003) *The Convention on the Sustainable Management of Lake Tanganyika*, Republic of Burundi, Democratic Republic of Congo, United Republic of Tanzania and Republic of Zambia; available at ⟨http://www.ltbp.org/LGLCON.HTM⟩.
Mkawale, S. (2005) "Residents Barricade Road as River Dries up", *East African Standard* (Nairobi), 15 February; available at ⟨http://allafrica.com/stories/ 200502150713.html⟩.
Mwangi, M. (2005) "Fighting Flares up Yet Again in Maai Mahiu", *The Nation* (Nairobi), 19 February; available at ⟨http://allafrica.com/stories/200502180779. html⟩.
OAS [Organization of American States] (2000) *Inter-American Strategy for the Promotion of Public Participation in Decision Making for Sustainable Development*, ⟨http://www.ispnet.org/Documents⟩.
Parr, J. (2005) "Protect Natural Barriers to Cut Tsunami Impact", *Asahi Shimbun*, English edn (Tokyo), 22 February, p. 26.
Sklarew, D., S. Annis, J. R. Mendler and M. Hamid (2001) "Forging a Global Community to Address International Waters Crises", *Water Resources Impacts*, April; available at ⟨http://www.iwlearn.org/ftp/iwl-in-wri.pdf⟩.
South Africa (1996) *Constitution of the Republic of SA (Act No. 108 of 1996)*, ⟨http://www.info.gov.za/documents/constitution/1996/96cons2.htm⟩.
SRH [Secretaria de Recursos Hídricos do Ministério do Meio Ambiente, Brasil] (2003) "SRH prepara proposta para o Sede Zero" ("Water Resources Secretary Prepares Proposal for Zero Thirst"), *Noticias* (Brazilia), 27 May; available at ⟨http://www.mma.gov.br/ascom/ultimas/index.cfm?id=391⟩.
Suarez, P. and D. Sklarew, eds (2002) "Transboundary Waters Management: Perspectives from Latin America and Caribbean Managers", *Proceedings of the First International Symposium on Transboundary Waters Management*, Monterrey, Mexico, November; available at ⟨http://www.iwlearn.org/ftp/GEF-IW-LAC-2001-EN.pdf⟩.
TAI [The Access Initiative] (n.d.) "Research and Results", ⟨http://www. accessinitiative.org/results_and_findings.html⟩.
UNDP [United Nations Development Programme] (1997) *Empowering People – A Guide to Participation*. New York: UNDP; available at ⟨http://www.undp.org/ csopp/CSO/NewFiles/docemppeople.html⟩.
UNECA [United Nations Economic Commission for Africa] (1990) *African Charter for Popular Participation in Development and Transformation*. International Conference on Popular Participation in the Recovery and Development Process in Africa, Arusha, Tanzania, 16 February; available at ⟨http://www. uneca.org/eca_resources/Publications/DMD/enhancing_african_csos/annex1. doc⟩.
UNECE [United Nations Economic Commission for Europe] (1992) *Convention on the Protection and Use of Transboundary Watercourses and International Lakes, Mar. 17, 1992* [Helsinki Convention]; available at ⟨http://www.unece. org/env/pp/documents/cep43e.pdf⟩.

——— (1998) *Convention on Access to Information, Public Participation in Decision-Making and Access to Justice in Environmental Matters, June 25, 1998* [Aarhus Convention]; available at ⟨http://www.unece.org/env/pp/documents/cep43e.pdf⟩.

United Nations (1992) *Rio Declaration on Environment and Development*, Rio de Janeiro, Brazil, 13–14 June; available at ⟨http://www.un.org/documents/ga/conf151/aconf15126–1annex1.htm⟩.

——— (2000) *Resolution Adopted by the General Assembly, United Nations Millennium Declaration*, 55/2. New York, 8 September; available at ⟨http://www.un.org/millennium⟩.

United States Census Bureau (2005) *World POPClock Projection – Monthly World Population Figures (for January 1 2005)*. Washington, DC: US Census; available at ⟨http://www.census.gov/cgi-bin/ipc/popclockw⟩.

UPTW [Universities Partnership for Transboundary Waters] (2003) *Stakeholder Participation in International River Basins: Models, Successes and Failures*, Workshop. Corvallis, OR, 14–16 April; available at ⟨http://waterpartners.geo.orst.edu/new.html#past⟩.

Vannuccini, S. (2003) *Overview of Fish Production, Utilization, Consumption and Trade, Based on 2001 Data*. Rome: Food and Agriculture Organization Fishery Information, Data and Statistics Division; available at ⟨ftp://ftp.fao.org/fi/stat/overview/2001/commodit/2001fisheryoverview.pdf⟩.

WCPFC [Western and Central Pacific Fisheries Commission] (2000) *Convention on the Conservation and Management of Highly Migratory Fish Stocks in the Western and Central Pacific Ocean*; available at ⟨http://www.ocean-affairs.com/pdf/text.pdf⟩.

WHO [World Health Organization] (2003) *The Right to Water*. Geneva: WHO; available at ⟨http://www.who.int/water_sanitation_health/rightowater⟩.

Wikipedia (n.d.) *International Waters*. St. Petersburg, FL: Wikimedia Foundation; available at ⟨http://en.wikipedia.org/wiki/International_waters⟩.

World Bank (2000) *Involving Nongovernmental Organizations in Bank-Supported Activities*. World Bank Operational Manual, Good Practice statement GP 14.70. Washington, DC: World Bank, February; available at ⟨http://wbln0018.worldbank.org/Institutional/Manuals/OpManual.nsf/9f854ba8ce7d9b85852565af0054aa88/1dfb2471de05bf9a8525672c007d0950?OpenDocument⟩.

——— (2004) *Environmental Assessment*. World Bank Operational Manual, Operational Policy OP 4.01. Washington, DC: World Bank, August.

WSSD [World Summit on Sustainable Development] (2002) *Johannesburg Plan of Implementation (Chapter IV)*. Johannesburg, South Africa: United Nations Department for Economic and Social Affairs; available at ⟨http://www.un.org/esa/sustdev/documents/WSSD_POI_PD/English/POIChapter4.htm⟩.

7

Development of an email-based field data collection system for environmental assessment

Srikantha Herath, Nguyen Hoa Binh, Venkatesh Raghavan, Hoang Minh Hien, Nguyen Dinh Hoa, Nguyen Truong Xuan

Introduction

Sustainable development principles call for a development process that does not over-utilize resources and compromise the needs of future generations. This translates into managing and preserving the environment, with the public acting as stewards of the environment. This requires effective means for public participation in environmental management and decision-making. How can this be achieved?

Currently, environmental impact assessment (EIA) provides a mechanism for the public to voice concerns over proposed development activities that might have an adverse impact on communities. Although EIA is effective and is being systematically used in assessing the impacts of single projects or acts, it is not adequate for addressing the whole spectrum of issues surrounding environmental management. Furthermore, it is a reactive approach and is limited to particular aspects of a proposed development plan, rather than being involved in the management of the environment more generally. Cumulative impact assessment looks at the cumulative effects of existing and anticipated stresses, in addition to the stresses that may come about as a result of the anticipated stresses from a planned activity. Regional impact assessment recognizes the activities and impacts within a region that could be interconnected, as in climate change impacts and transboundary basin issues. It is not sufficient to consider only local actions and impacts; actions committed far away that would have an impact locally, as well as the impacts of actions conducted locally on interconnected far-away places, need to be addressed. Another level of complexity is added when strategic environmental assessment is brought in, where one examines the impacts of policies, plans and programmes.

It is clear that broadening the horizon of environmental assessment increases the complexity of the issues as well as the scope and depth of the issues to be covered. Therefore environmental assessment becomes closely associated with information collection, analysis and management. This can be achieved only through the development of better procedures and tools that can enable the participation of a broad segment of society in environmental management.

Recent trends clearly show a movement towards increased public participation in environmental management (UNEP 2000). Many programmes dealing with public participation in environmental management are spearheaded by non-governmental organizations and communities affected by changes to the environment brought about by various development programmes. It is also recognized that national environmental policies can be more effectively implemented if they are supported by an informed public that has had the opportunity to participate in the management of the environment. Towards achieving this, environmental education and awareness programmes are expanding at all levels and in all sectors. These investments in education will result in a population that is more knowledgeable about the environment and eager to participate in its management. It is necessary, then, to develop procedures to facilitate public participation to enlist this growing support base for environmental management.

In order to manage the environment, it is necessary to know its current state, how it would change under different resource utilization scenarios, and the effect of these changes on the communities linked with the environment. As mentioned already, this means knowledge of the magnitudes and interrelations of physical environment parameters, their distribution in space and time, human activities and their impact on the environment, and vice versa. It is important to recognize that a large amount of information-gathering and synthesis is required to achieve this. Although information on environmental resources such as the amount of water in a lake or groundwater levels would indicate the current status, it is necessary to develop models that relate these states with climatic input and human interventions. Because a model is only an idealized representation of reality, it is necessary to validate the models' predictions with observations on the ground before the models are used to predict future conditions.

The nature of the environmental crisis brings further complexity. Most environmental changes occur very slowly: changes between weeks, months or sometimes even years are not very significant. However, the accumulated changes over a longer period of time could increase the vulnerability of a community and suddenly bring about a severe crisis situation. Environmental assessment should therefore be a continuous process that has provisions to accommodate parameters that describe the environment as a dynamic evolving entity.

One of the major difficulties in current approaches to environmental modelling lies in the acquisition of spatial information of sufficiently high resolution and accuracy to describe and forecast impacts resulting from the changes. Although remote sensing techniques have made big advances in the recent past in terms of coverage and resolution, the information they can provide is still limited to a few parameters, such as land cover, and accuracy is constrained by the amount of data directly measured on the ground available for the calibration of remotely sensed data. The information available on many other parameters, such as groundwater, surface water conditions and irrigation water use, remains incomplete. Data on the environment, and their distribution in time and space, are especially lacking in the developing countries, where governments are able to allocate only limited resources for data acquisition.

Advances in information technology (IT) have now made it possible to communicate and disseminate information easily. The procedures and tools discussed above are best implemented using IT tools in order to facilitate the wider participation of stakeholders. New developments in technologies should be used to create tools for community needs because otherwise these technologies would be of marginal use. At the same time, if these new technologies are not used in the environmental decision-making process, the power and opportunities they bring would be wasted (Craig et al. 2002). In this chapter, we describe the development of an email-based data acquisition system to address some of the above concerns. An email-based system is ideal for developing countries where low bandwidths still prevent wide usage of the Internet, whereas email is widely used for communication and information-sharing. The system has the advantage of enabling local communities to provide information on environmental management because local citizens are often involved in activities that lead to environmental change, and in addition they have the most intimate knowledge of local conditions (Brosius et al. 1998).

Background

Typical centralized data compilation and maintenance systems invariably follow a standard procedure. First, field surveys are carried out and the variables are recorded in standard forms, which are then collected in regional offices where they are computerized and sent to a central office. After verification, the data are inputted to a database application, which tends to be specifically developed for the storage and retrieval of that particular set of information to meet a set of predefined needs. In general, such systems are very specific and are not updated regularly with hardware and software upgrades and cannot provide services to a large user base. On the other hand, recent developments in Internet and associated information dissemination technologies provide many methods for transparent data storage and seamless integration with other applications. Further, these solutions are generally scalable and can accommodate growing data needs. However, use of such technologies in developing countries is still problematic owing to low Internet accessibility and slow connection speeds. In most developing countries, the available bandwidth is limited and the Internet is primarily used for text-based information retrieval or for email only. Under such conditions, it is necessary to develop special applications to utilize data collection and dissemination facilities based on technologies made possible by the rapid growth of the Internet for data management.

The email-based data collection system described here was developed as a collaborative effort in partnership with the Media Center of Osaka City University, Japan, the Center for Information Technology of Hanoi University of Mining and Geology, the Institute of Information Technology of Vietnam National University, Hanoi, and PeaceSoft Solutions Co., Hanoi, Viet Nam. Viet Nam was chosen as the venue for development, so that the system could be developed and tested in an environment that experiences actual limitations on Internet access.

System description

The overview of the system is shown in Figure 7.1. The two main components are the client-side functions, which send data to a central server, and the server-side functions, which update and maintain a database as well as communicate with the users. A relational database structure is used for storage and retrieval to facilitate efficient transactions, updates and retrieval (Ullman 1988). Communication between the user input and the database is accomplished using an XML to database translation model (Florescu and Kossmann 1999). The client-side components provide facilities for users to fill in an XML template that describes the data and the values. The template is submitted to the database server via an email. On the server side, a mail robot program is configured to receive user mails and respond to users. Whenever a new email comes in, an email gateway program is automatically called to parse the email

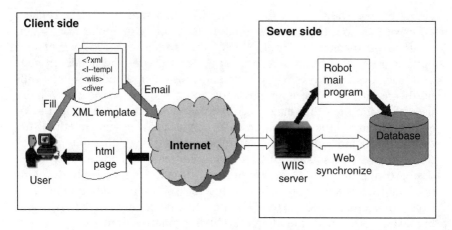

Figure 7.1 System overview.

and add the information received to the database (Linden 1999). A web server is used to synchronize data in the database and to provide a web interface for users to view already archived data, add new data or update the database using a web browser. The details of client-side handling are shown in Figure 7.2. The data entered by the user are parsed into an XML file, corresponding to the structure of the database table where the data will be finally stored. Images from sites can be attached and sent by any mail client, addressed to the database server.

Figure 7.3 shows details of the server side. The database server functions at two levels, as shown in the figure. As a web server, it provides facilities for users to access, browse, search and download data from the database. On the mail server side, a mail program parses the received email and the data are added to the corresponding data table. An XML parser is prepared in this application to check the validity of the data.

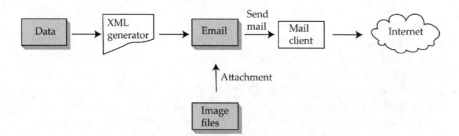

Figure 7.2 Details of client side.

AN EMAIL-BASED FIELD DATA COLLECTION SYSTEM

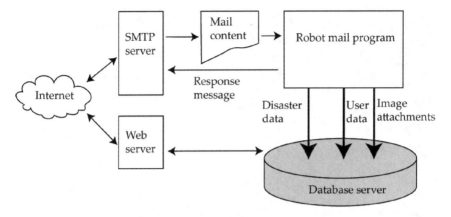

Figure 7.3 Server-side details.

The user is acknowledged by an automatic message, which also sends notification if any data inconsistencies are identified by the parser.

System use

The system described can be used for various types of information-gathering. The client XML generator as well as the server-side XML parser can be configured to handle different types of data structures stored in different data tables. The user is supplied with a simple menu-based program to input data, which are automatically formatted with XML tags. For example, to collect information on different types of river water use, a template as shown in Figure 7.4 is provided. The application takes user information and generates the XML code, which is inserted into an email. Similarly, different data structures can be employed to collect different types of information, such as those related to disaster loss estimation or building characteristics. For each type of information, a simple client-side form is available for data input, which converts the data to an XML tagged email that is sent to the server. The XML parser identifies the type of data and automatically selects the database tables to be updated. The web server provides facilities to list the data tables available in the system as well as to query and download data for general users. In addition, editing, inserting and deleting facilities are provided to administrators.

The system has been field tested and found to be easy to use in field conditions, where data transfers using mobile phones also worked well without any difficulties.

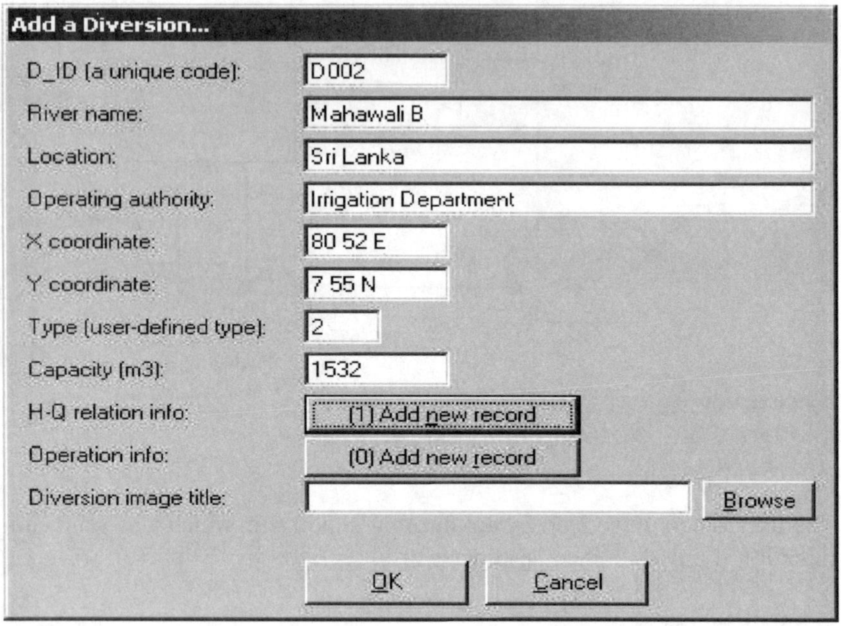

Figure 7.4 Input data form for the user.

System implementation

The system has been developed entirely using open-source software applications to ensure that it can be easily implemented by different organizations interested in data archiving and dissemination. Red Hat Linux was selected as the operating system. The particular choice was made to ensure easy transfer and installation procedures, given the wide use of this particular source code arrangement. For the mail server, Postfix is used. Postfix is available in all Linux platforms and supports SMTP and POP3 for email transactions. For the Relational Data Base Management System (RDBMS), either the MySQL or the PostgressSQL database can be used. The particular choice will depend on each implementation, especially if additional connectivity to services such as geographical information systems is contemplated.

The system configuration is carried out in a few easy steps. First the Postfix SMTP server is configured to communicate with the robot mail program. Several modules are defined to achieve the following functionality. The incoming mail, formatted in XML, is received and the relevant data are extracted. The database stores these data in a temporary table and generates a report containing the data stored; then the report is emailed back to the sender for confirmation. This gives an opportunity for the person who made the submission to verify the data and correct them if necessary. The confirmed data are then moved to the permanent data storage area.

The system can be used to handle different types of data for different applications. For example, an application on agricultural productivity may be concerned with cultivation dates, cultivation parameters, time history, location and the extent of agricultural land and fertilizer applications, whereas land cover information for environmental studies may be concerned with geographical information related to the extent of land classes, land classification, physical characteristics such as leaf area index, root distribution, soil classes and density. Different data tables are predefined for each such application. If additional data need to be collected for an existing application, an additional data table is created to accommodate the need and linked with the main data table.

To submit data to the system, a small utility program is provided for each application. The application generates the XML template in which the user types and sends the data to the system. Once data are input, the utility program automatically generates the XML tagged email to be sent to the database mail server. One of the drawbacks of this arrangement is the need to replace the client-side program with database changes or expansions. In the future this should be addressed by providing a mechanism to detect the client program version and to provide upgrades from a repository of utilities corresponding to the different types of information being collected.

Application to water infrastructure information collection

Physically based hydrological modelling has now emerged as the most rational method for assessing water conditions for efficient water management as well as for real-time applications utilizing the most up-to-date information pertaining to catchments (Herath et al. 1999; Dutta and Herath 1999). The accuracy of physically based approaches directly depends on the accuracy and completeness of the physical data used in the modelling and assessment. In addition to physical information such as elevation and land cover, water utilization and storage practices in a river basin are extremely important for understanding and forecasting basin hydrology. These human interventions are carried out through various types of infrastructure. Although their impact on the water circulation in a catchment is very high, information pertaining to them is often incomplete and unreliable. In the present study, we aim at implementing a generalized database for such infrastructures that can be filled with data as well as accessed through the Internet. The infrastructure is identified by type, i.e. retention or conveyance. Retention-type infrastructures are described by location, category, amount of water, operational policy and capacity. The conveyance structures are described by starting and ending locations, purpose, operational attributes, and physical characteristics. Additionally, cross-sections, water level at different flows and roughness at different locations can be linked to the conveyance structure information.

The database structure of the system running on the central server is shown in Figure 7.5. A sample record entry form used to fill in data related to reservoir information sent from Sri Lanka has already been illustrated in Figure 7.4. The record inserted on the central system is shown in Figure 7.6, along with other existing data. Finally, Figure 7.7 shows the list of infrastructure information that is available in the system and accessible over the Internet. Each entry is linked with retrieval functions that

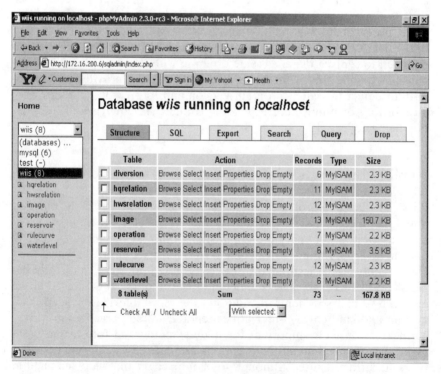

Figure 7.5 The water infrastructure information system (WIIS) data structure.

AN EMAIL-BASED FIELD DATA COLLECTION SYSTEM 129

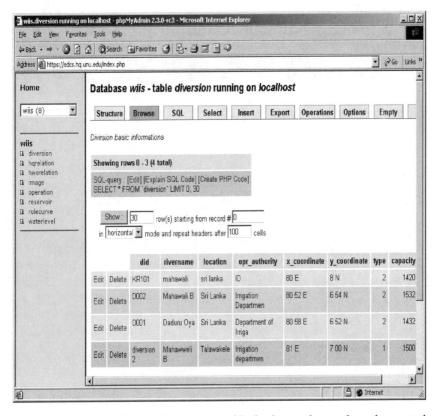

Figure 7.6 List of existing infrastructure with the inserted record on the central database.

can generate either full reports for the infrastructure or specific information related to characteristics of the infrastructure.

Potential uses of the system

The email-driven system was developed to collect spatially distributed information and has been tested in a water infrastructure data collection application. The system can be used for various other purposes by creating data tables in the system to match the data structures required by the applications. Its low bandwidth requirements and open-source usage make the system especially attractive to developing countries. Given the

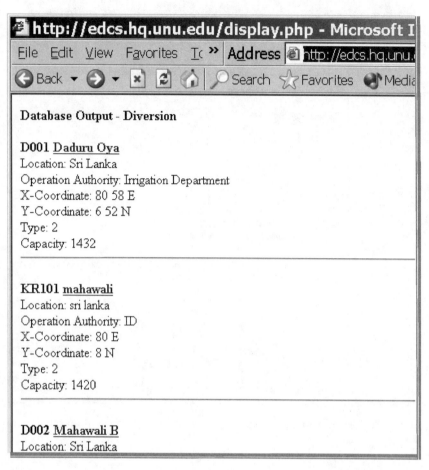

Figure 7.7 A sample list of database entries in the water infrastructure information system.

wide availability of email facilities, it is recognized that the system can be used easily from the field using different types of communication tools.

The familiarity of email usage makes it possible to expand the user base from specialized users to the general public, who can play a very important role by using such systems to help improve spatial data sets on infrastructure as well as for environmental monitoring. The present planned extensions to the program include linking with geographical information systems and automated report generation mechanisms.

This implementation has demonstrated the feasibility of developing appropriate IT tools for collecting and managing cost-effective databases.

The development in partnership with institutions in the developing countries has proved to be valuable in identifying an appropriate implementation platform and software. It also ensured that the resulting system can be used in the field effectively, and, more importantly, that it can be maintained and further improved by any interested party with very little investment.

At present the main users are envisaged to be personnel responsible and authorized to collect and disseminate data related to the environment in different countries, especially in transboundary basins where often a complete understanding of basin hydrology is not available owing to the lack of communication and data exchange mechanisms. A basic understanding of a basin water cycle is fundamental to its management under different development plans, and a common platform for information-gathering and information-sharing for analysis is a prerequisite to building a consensus on these issues. The email-based data collection system could be an ideal supporting tool for such an on-line decision support system, for example a system that simulates the state of hydrology in a transboundary basin. Although the baseline data provision can be restricted to authorized providers, information from the general public on impacts and observations related to past experiences can easily be solicited by providing facilities to update existing information, provide new data and retrieve forecasts, which would also help improve the predictability as well as the verification of such simulation systems. Furthermore, the participation by various stakeholders could be promoted by providing facilities for carrying out future scenario analysis through an on-line decision support system using data sets provided by the email system described above.

REFERENCES

Brosius, P. J., A. Tsing and C. Zerner (1998) "Representing Communities: Histories and Politics of Community-based Natural Resource Management", *Society and Natural Resources* 11: 157–168.

Craig, W., T. Harris and D. Weiner (2002) *Community Participation and Geographic Information Systems*. London: Taylor & Francis.

Dutta, D. and S. Herath (1999) "Methodology for Flood Damage Assessment Using GIS and Distributed Hydrologic Model", in *Proc. International Symposium on Information Technology Tools for Natural Disaster Risk Management*, INCEDE Report 11, University of Tokyo, pp. 109–124. Bangkok: INCEDE–UNU–AIT.

Florescu, D. and D. Kossmann (1999) "Storing and Querying XML Data Using an RDBMS", *IEEE Data Engineering Bulletin* 22(3): 27–34.

Herath, S., R. Jha, D. Yang and S. Dutta (1999) "Preparation of Spatial Data Sets for Physically Based Hydrological Models", *Proc. International Symposium on Information Technology Tools for Natural Disaster Risk Management*, INCEDE Report 11, University of Tokyo, pp. 199–214. Bangkok: INCEDE–UNU–AIT.

Linden, T. (1999), htnews documentation, ⟨http://www.co.daemon.de/en/software/htnews/⟩.

Ullman, J. D. (1988) *Principles of Database and Knowledge-base Systems, Volume I*. Rockville, MD: Computer Science Press.

UNEP [United Nations Environment Programme] (2000) *Global Environment Outlook 3: Past, Present and Future Perspectives*. London: Earthscan.

8
New directions in the development of decision support systems for water resources management

Kazimierz A. Salewicz

Introduction

The decision-making processes associated with a broadly understood utilization of natural resources and water management fall into the category of complex situations requiring very thorough consideration and analysis. The complexity manifests itself not only through the sophistication of physical, chemical and biological processes taking place in water resources and land-use systems, but primarily through very rich and multidimensional interactions between various types of more or less thought-out human activities, their influence on natural systems and the impact on the human world resulting from the responses of these natural systems.

Owing to the intrinsic complexity and sophistication of the decision problems associated with water resources management, the relevant decision-making processes cannot be implemented without making use of modern technological means, especially computer technology. In this chapter, I first briefly discuss the main trends underlying the development of tools used to support decision-making processes. Then, based upon progress made in the area of Internet technology, I present a Web-based prototype of the decision support system that has been developed for the Ganges River. The experiences and lessons learned during the development of this prototype system allow me to discuss the directions of possible further developments.

Decision problems in natural and water resources management

The decision-making associated with the utilization of natural and water resources is understood here as the process of selecting those actions influencing the behaviour of a given natural resources system that at least in theory (to allow for various false or misinformed decisions, which happen all the time throughout the world) should result in a better fulfilment of the goals and objectives by the system under consideration. Decision-making can be also understood as a process of seeking the "best acceptable" solution for a specific system.

Decision-making processes take place in a general structure consisting of the following elements (see Figure 8.1):
- the **system** (for instance, a water management system) under consideration, representing the material and physical reality;
- the **problem** requiring a decision – the "problem" is the existence of a gap between the desired state and the existing state (Sabherwal and

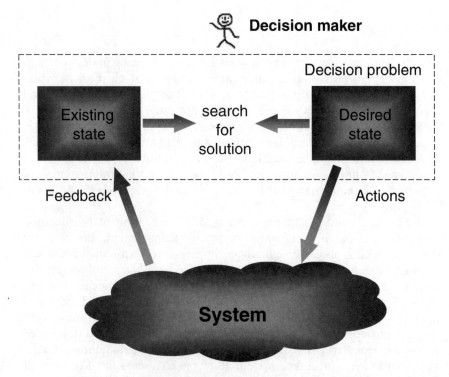

Figure 8.1 Components of the decision-making process.

Grover 1989), which the decision-making process aims to fill, or at least reduce, and thus solve the problem; and
- the **decision maker**, the person or personalized organization who is required to decide upon the action or a set of actions to be undertaken in order to achieve certain objectives (e.g. to fill or reduce the gap between the existing and the desired state of the system). These objectives are provided by those to whom the decision maker is responsible. Most methodologies assume an individual decision maker. However, in real-world situations, decisions are usually made by a group or even groups of people representing different views, preferences, expectations, etc.

Despite the multiplicity of examples of decision problems and approaches to solving them, one can distinguish two basic categories of decision-making problem that are typical of the management of human-made systems, but primarily concern water resources management in both a national and an international context:
- structural decisions associated with considering the development and/or removal of structural components of the system (such as building a new dam, constructing a new water treatment station, disassembling a nuclear power plant);
- operational decisions concerning how already existing (or planned) elements of the system should be operated.

Structural decisions concern changes to the topological and/or geographical structure of water resources systems and are – generally speaking – associated with the implementation of engineering schemes. The time horizons necessary to consider and evaluate the consequences of these decisions are usually very long, extending over many years, if not centuries.

Operational decisions are based on the assumption that the topological and engineering structure of the water management system is given (fixed) and the essence of the decision-making problem is the question of how to make the best use of the already existing structure and its components to fulfil the operational objectives. The decision problems related to the search for operational policies are usually solved over shorter time horizons, varying from minutes or hours (for hydropower generation and flood control) to weeks and months (for retention control).

Both types of decision-making problem are characterized by:
- uncertainty concerning the future state of nature, the environment, human requirements and expectations;
- imprecise and sometimes unknown conflicting and multiple objectives;
- a high degree of nonlinearity and dynamics.

These characteristics have a very strong impact on structuring and organizing the process of public participation in the management of natural resources. Consequently, they also very strongly influence the development of tools capable of supporting decision-making and enabling active and constructive public participation: the concepts underlying the development of these tools have to address the above-mentioned characteristics directly and fulfil the requirements imposed by the need to ensure the applicability of these tools in a decision process involving public participation. Systems-based approaches developed during recent years have allowed us to investigate, analyse and model complex interactions, gradually preparing the ground for the search for development strategies and operational policies for various types of water resources system.

The technical and technological developments taking place in the domain of systems analysis and information technology have led to significant progress and advances in the fields of hydrology, water resources management, and environmental and decision sciences. For decades the evolutionary process associated with the development of models and tools for water resources management has also very closely reflected progress in the domain of mathematical modelling, linear and non-linear optimization, stochastic modelling, programming languages and data-processing.

This significant progress is extensively documented in a very rich literature. It is evident in the creation of various approaches and tools. However, even some of the models and tools created recently have been built upon still valid "traditional" notions and concepts underlying the operation of water resources management and multiple reservoirs systems, such as storage zones and rule curves (Loucks and Sigvaldason 1982), which were developed several years ago. The fundamental work in the area of storage reservoir operation and flow routing was done at the Hydrologic Engineering Center (HEC) of the US Army Corps of Engineers (USACE) at Davis, California, where in the course of several decades a number of models and tools have been developed, such as:
- HEC-1: flood hydrograph package;
- HEC-2: water surface profiles model (USACE 1992);
- HEC-3: reservoir systems analysis model (USACE 1985);
- HEC-5: reservoir operation simulation model containing water quality components (USACE 1982 and 1986);
- HEC-RAS: river analysis system containing graphical information systems extensions (USACE 1995).

In the area of environmental processes, traditional simulation approaches resulted in a number of models describing the complex physical and chemical phenomena taking place in water bodies. For example, Li and Chen (1994) developed a model for simulating the removal of organic matter and oxygen consumption by biofilms in an open channel. A probabilistic method for uncertainty analysis and parameter estimation for dissolved oxygen models has been developed by Masliev and Somlyody (1994). A numerical model for water quality simulation has been developed by Kazmi and Hansen (1997) and applied to a case study in the Yamuna River in India.

In addition to simulation, optimization techniques have been widely used in the field of water resources management, environmental management and pollution control. The results obtained from solving various types of optimization problem have provided the basis for making decisions related to reservoir operation, hydropower generation, groundwater management, the allocation of waste loadings and the implementation of pollution abatement activities, and many others. Examples of relevant publications include works by Allan and Bridgeman (1986), Marino and Loaiciga (1985), Kelman et al. (1989), Ellis and ReVelle (1988), Wang and Zheng (1998), Chen and Chang (1998) and Ning and Chang (2002). Multi-objective techniques and evaluation have received wide attention in environmental applications (for example, solid waste management), as reported by Minor and Jacobs (1994). Further steps have been made by researchers such as Chang et al. (1997) and Chang and Wei (1999), who reported the development of an approach for routing and scheduling collection vehicles in a solid waste management system by combining multi-objective, mixed-integer programming with geographical information systems (GIS).

Decision problems are handled and solved in complex structures and processes involving many stakeholders representing various groups, groups of interests, political orientations, etc. Depending upon the concrete decision situation, the information requirements and needs expressed and/or perceived by the stakeholders in the decision-making process can be very different. Experience shows that it is impossible to specify beforehand what information will be necessary and sufficient to make good decisions. Usually the decision-making process goes hand-in-hand with a learning process. In the framework of the learning process, the stakeholders make decisions based on the information available; they learn about the impacts and consequences of those decisions and then make further decisions in the light of the new knowledge and information they have acquired. Thus, in a repeatable process they enhance their knowledge and understanding of the decision problem and also identify needs for new types of information. Information needs and requirements therefore grow together with the growing understanding of the problem at hand.

An interesting discussion concerning this subject is provided by Simonovic (2000) in the context of a complexity paradigm relevant to water problems. Population growth, climate variability and regulatory requirements are increasing the complexity of water resources problems. Water resources management schemes are planned for longer temporal scales in order to take into account and satisfy future needs. Planning over longer time horizons extends the spatial scale as well. The extension of temporal and spatial scales leads to greater complexity in the decision-making processes and involves increasing the number of stakeholders.

Consequently, together with the growing complexity of the decision-making problems, there are growing demands and challenges concerning the tools used to provide information and to support the decision-making processes. The complexity associated with the utilization and management of water resources calls for tools capable of mirroring the complexity of the problems under consideration. At the same time, these tools have to be able to cope efficiently with the multiplicity and volume of information that has to be processed during decision-making. The ability to process relevant information must be accompanied by the ability to present this information to the user and, consequently, to a decision maker. These capabilities are provided by decision support systems.

Decision support systems – introduction

Decision support systems can be defined as computer technology solutions that can be used to support complex decision-making and problem-solving (see Shim et al. 2002). Although this definition applies very well to decision-making in many purely technical areas, it fails to reflect one extremely important aspect of the decision-making process in water resources systems: the role of the human factor.

Owing to the very complex nature of water resources management problems, the lack of consistent and complete data, uncertainties, and the poorly structured form of decision problems, the process of finding decisions cannot be limited to solving mathematical optimization problems or performing complex simulations. In this context, we understand the decision support system (DSS) as a set of computer-based tools that provide decision makers with interactive capabilities to enhance their understanding and information basis of the decision problem under consideration through the use of models and data-processing, which in turn allows decisions to be reached by combining personal judgement with the information provided by these tools.

DSS was born in the early 1970s and evolved from two main areas of research: the theoretical studies of organizational decision-making conducted at the Carnegie Institute of Technology during the late 1950s; and the technical investigations carried out at the Massachusetts Institute of Technology in the 1960s (see Keen and Morton 1978).

THE DEVELOPMENT OF DECISION SUPPORT SYSTEMS

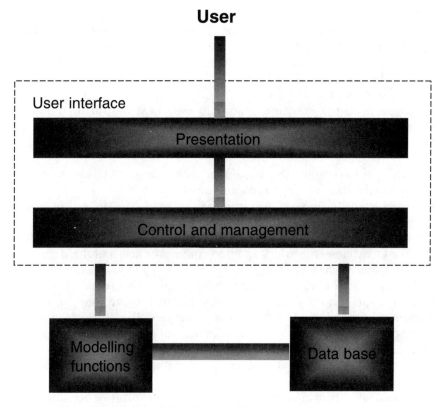

Figure 8.2 Main building blocks of the decision support system.

The classic DSS tool design, as shown in Figure 8.2, comprises the components for:
- database management capabilities with access to internal and external data, information and knowledge;
- powerful modelling functions accessed by a model management system; and
- user interface designs that enable interactive queries, reporting and graphic functions.

This view of decision support systems concerns their technical architecture and the building blocks that have to be incorporated into the design and development of DSS.

Over the past three decades, the developers and users of DSS have used broader or narrower definitions, and other solutions that do not fully meet the components listed above have also emerged to assist decision makers faced with specific kinds of problem. Nevertheless, the classic DSS architecture contains these three basic components.

Another, complementary way of looking at DSS is associated with the role and functions that DSS has to fulfil (Parker and Al-Utaibi 1986), as seen from the user's perspective:
- it assists managers in their decision processes in semi-structured tasks;
- it supports and enhances rather than replaces managerial judgement;
- it improves the effectiveness of decision-making rather than its efficiency;
- it attempts to combine the use of models or analytical techniques with traditional data access and retrieval functions;
- it specifically focuses on features that make it easy to use by non-computer people in an interactive mode;
- it emphasizes the flexibility and adaptability to accommodate changes in the environment in which the decision maker acts and the decision-making approach of the user.

The capability of the DSS to fulfil these functions is particularly important for its practical usability and acceptance by a broad range of stakeholders involved in the decision-making processes. The degree to which a specific DSS meets these characteristics and capabilities has a direct impact on its abilities to satisfy the information needs of the decision makers as well as the stakeholders participating in a decision-making process.

Traditionally, mathematical models and various forms of decision support tools and systems incorporating these models have been developed by analysts and modellers for the same type of audience. Therefore it was not necessary to pay any special attention to the design and implementation of user-friendly interfaces between the tool and its user. This situation continued for years and contributed, in fact, to the creation and growth of a gap between modellers and analysts, on one side, and decision makers (not to mention the general public), on the other side. As long as decisions were taken by a narrow circle of specialists, awareness of this gap was not so dominating and was not perceived as a meaningful factor limiting communication between stakeholders.

The situation became much more complicated when these tools were no longer used only by a limited range of modellers and analysts, and other, less technically minded and less technically experienced groups of users emerged and voiced the request, the right and the will to use these tools to secure active and informed participation in the decision-making process. Consequently, one of the biggest current challenges of the DSS in facilitating access to information by a broad spectrum of stakeholders is the fact that the available information must directly address their concerns and information needs. Another challenge is associated with the necessity of providing groups of technical non-professionals with the possibility of obtaining answers to questions that are important and meaningful to these groups, especially when neither the questions nor the responses necessarily have to be expressed in technical terms. The information presented to non-specialists cannot substitute for or hide real facts. This information must have the same value as far as the actual consequences of considered decision alternatives are concerned, but the form of this information should allow for straightforward recognition of the impacts, dangers and benefits.

A common and broadly applied development approach to address the expectations and needs of a broad range of stakeholders in decision problems is to secure a potentially high degree of flexibility through the interactive and graphics-driven operation of decision support systems. Here I present some representative examples of decision support systems offering their users various levels of interactivity, computational capabilities and sophistication:

- The Interactive River Simulation (IRIS) system developed in the late 1980s (see Loucks and Salewicz 1989; Loucks et al. 1990). It was developed with the intention of being used as a decision support and alternative screening tool for assisting decision makers and stakeholders involved in resolving conflicts associated with the management of international river basins (see Salewicz and Loucks 1989; Venema and Schiller 1995; Salewicz 2003). An extended and improved version of the model – the Interactive River-Aquifer Simulation (IRAS) programme (see Loucks and Bain 2002) – has been developed primarily to assist those interested in evaluating the performance of watershed or regional water resources systems.
- ModSim, a general-purpose river and reservoir operation simulation model, was originally developed by J. Labadie of Colorado State University in the mid-1970s (see Labadie 1995; Fredericks et al. 1998; Department of Civil Engineering, Colorado State University 2000; US Department of the Interior 2000) to enable the simulation of large-scale, complex water resources systems, including consideration of water rights, reservoir operation, and institutional and legal factors that affect river basin planning processes.
- RiverWare (see Zagona et al. 1998, 2001) represents a completely new generation of tools for the planning and management of river basin systems. The capabilities offered by "object-oriented" technology (see Booch 1994) allow for the development of new software by combining general modelling tools (classes, objects) that are not specifically designed for river basin systems within one modelling framework. RiverWare, developed at the Center for Advanced Decision Support for

Water and Environmental Systems of the University of Colorado in cooperation with the US Bureau of Reclamation, utilizes object-oriented software technology to create a flexible modelling framework by combining building blocks that describe the possible physical components of a water management system with specific solution procedures capable of tackling operational problems through simulation and/or optimization.

Other examples of implementations of decision support systems are provided by Andreu et al. (1996), Ford and Killen (1995), Hipel et al. (1997) and Ito et al. (2001).[1]

Progress and evolution in the area of decision support systems have been very strongly influenced and shaped by technological changes leading to rapid growth in computing capabilities. A useful metric for the rate of technological change is the average period during which speed or capacity either doubles or – more or less equivalently – halves in price (see Foster 2002). For computing power, storage and networks, these periods are around 18, 12 and 9 months respectively. The evolutionary progress resulting from technological change has brought with it transformations and advances in DSS primarily with respect to:

- increasing level of interaction with the user (from batch processing to interactive operation);
- growing complexity and detail of coverage of the phenomena modelled;
- rising availability of computing equipment and decision support tools accompanied by simplified access to these instruments by growing circles of users (from access limited to specialists to public access, which is so important for public participation in decision-making processes);
- broad usage of comprehensive data (from point-related data to geographical information systems and distributed databases).

Entirely new technical and conceptual possibilities for the further development of DSS have been created by revolutionary changes in Internet technologies. At first, these allowed for access to knowledge and information (document retrieval); now they support data generation and retrieval using Web services, which are run on hardware remote from the user and are managed by a service provider.

Prototype implementation of a Web-based decision support system

One of the main advantages of the Internet is its ability to provide almost unlimited access to information to anybody who is technically capable of connecting to the Web. This makes the Internet particularly suitable for providing and disseminating information relevant to decision-making processes associated with the utilization and management of water resources. Specifically, the Internet may be seen as a platform allowing for public participation in disputes and the resolution of conflicts, especially in a case involving many stakeholders and in an international context. In such situations, access to neutral and unbiased sources of information about the consequences of possible policy and/or management strategies may contribute to constructive dialogue between the parties involved and consequently help to overcome the differences that divide the parties and reach, if not an agreement, then at least a reasonable compromise. There is still a long way to go to full implementation of a dedicated Web-based DSS and it involves many political, organizational and technical difficulties. Nevertheless, this ultimate goal can be gradually achieved through stepwise efforts and the progress resulting from these efforts.

To explore the technical possibilities and feasibility of developing a decision support system using the Internet, I have undertaken research aimed at developing a prototype (pilot) installation of a DSS on the Web. The development of this prototype is based on the following assumptions concerning the characteristics of a target user and the underlying design principles:

- the prospective user of the DSS is a non-professional representative of the general public and is interested in assessing the consequences of a particular policy expressed in terms of clearly identified alternative actions;
- the actions associated with the policy are preferably formulated in a qualitative manner, and not quantitatively;
- the user has no experience of and no desire to learn about the specifics of any mathematical models and tools; moreover, the user has no experience in programming and wants to use the tool in an interactive manner via a graphical interface that is as simple as possible and does not require any programming;
- the tool should allow for the simple selection of available alternatives and present the consequences of the selected decisions in a meaningful way;
- the time interval between the formulation of the query and obtaining a response should be minimal.

The selection and formulation of the target user characteristics reflect the desire to develop a tool capable of facilitating public participation in discussing and consequently resolving controversial decision problems associated with the management of water resources. The extent to which public participation can be facilitated by this and similar decision support tools is very strongly associated with the following characteristics of the DSS:

- its broad availability to a possibly vast circle of users (which can be achieved by Internet access to the DSS);
- the intuitive and simple form of the tool, which allows it to be used without special training and education;
- the ability of the tool to provide the user with the possibility for individual selection of policy alternatives or decisions;
- the ability of the tool to allow any user to assess and evaluate the impact of the selected policy based on computed physical values;
- the capability of the tool to address problems relevant to the perceptions and expectations of the public.

These characteristics concern mainly technical and technology-related aspects associated with the use of DSS in resolving issues of public interest. The design and technical implementation of the prototype Internet-based DSS presented in this chapter were based on the principles and characteristics just outlined. The social, political and organizational aspects of the decision process involving public participation, although extremely interesting, were not directly addressed in the framework of the study reported here. Nevertheless, initial efforts were directed towards the selection of an appropriate case-study system, which:

- could potentially attract a significant audience;
- concerns important and controversial water management issues (possibly international) involving the conflicting objectives and interests of multiple parties;
- has been described using sound, verified and viable modelling techniques;
- has been analysed and modelled by objective, unbiased and independent specialists, who are not involved in the controversy.

A long search led to the selection of the Ganges River case study (see Figure 8.3), which has been the subject of extensive research at the Center for Spatial Information Science at the University of Tokyo (see Ministry of Land, Infrastructure and Transport 2001).

This case study concerns the analysis of the impact of agricultural and urbanization policies applied in India to the Ganges River. The agricultural and urban development policies chosen by India have a direct impact on the amount of water in the Ganges River flowing into Bangladesh. Taking into account the lack of cooperation between these two countries (see Biswas and Uitto 2001) and their mutual distrust, the availability of an unbiased, independently developed model and DSS capable of analysing the consequences of selected policy options could help both sides to establish common ground for the discussion and evaluation of the alternatives.

The relevant policies that could be applied in India involve the following decision variables:

THE DEVELOPMENT OF DECISION SUPPORT SYSTEMS 145

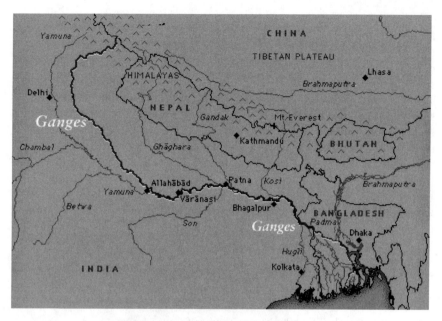

Figure 8.3 Map of the Ganges River basin.
Source: ⟨http://www.thewaterpage.com/ganges_map.htm⟩.
Note: © Microsoft Corporation. All Rights Reserved.

- the length of the stretch of river over which the agricultural and urbanization policies will be implemented;
- the intensity of the changes in land-use patterns; and
- the intensity of the urbanization changes in the area considered.

The policies can be described in detail in quantitative terms, using precise values for the above-mentioned decision variables and then the response of the system can be simulated for the selected values. However, one run of the simulation to calculate the response of the system to the selected policy alternative requires a couple of hours of computational time on very powerful machines (K. Rajan, personal communication, 2002). This property of the model could be seen as its disqualification, at least as far as the use of the model in an Internet-based, interactive decision support system is concerned. Therefore, instead of using an on-line simulation process to compute a system response to the decision variables selected by a user, a different solution had to be applied, namely off-line simulation of the system response to feasible policy alternatives. The results of these simulations are then stored in a separate database available to the user of a DSS.

Taking into account the fact that the average user of the model does not have enough knowledge and experience to experiment with the selection of precise numeric values for the decision variables, and also in order to limit the number of plausible policy alternatives for which off-line simulations had to be performed, we had to look for another approach. This approach is based on the concept of qualitative qualification of decision variables: the feasible range of every decision variable was divided into a small number of sub-intervals. All values of the decision variable in a certain sub-interval were associated with a single, qualitative attribute characterizing this range in descriptive terms (i.e. low, medium, high). This process of qualitative categorization of decision variables can be performed only on the basis of very thorough sensitivity analysis and expert knowledge of the models used to calculate the impact of policy parameters (K. Rajan, personal communication, 2002).

Following this concept, the feasible ranges of the decision variables expressed in descriptive terms were defined as follows (see Figure 8.4):

- The length of the area upstream of Harding Bridge where the changes to land-use policies were to be introduced was divided into three categories:
  - changes on the stretch shorter than 100 km;
  - changes on the stretch between 100 and 200 km; and
  - changes on the stretch longer than 200 km.
- The intensity of the change in land-use patterns was divided into four categories:
  - a shift in the cropping pattern to a more intensive pattern;
  - a shift from the current pattern to a less intensive one;
  - no change in the land-use pattern (retain current conditions); and
  - an increase in the irrigation command area, which is equivalent to the creation of bigger farms.
- The intensity of the urbanization changes in the area under consideration had three alternatives:
  - no change to the current density of the population;
  - an increase in the population density of up to 50 per cent; and
  - an increase in the density of up to 100 per cent.

The user who wants to see the impact of changes in land-use policy in India selects a combination of policy parameters expressed in descriptive terms as defined above.

The impact of the policy alternative may vary depending on the natural climatic conditions, characterized in this region of the world by the monsoon. Also in this case a qualitative description of climatic conditions was used: the impact of land-use policy is analysed using three alternative scenarios of climatic conditions extending over a one-year time horizon for:

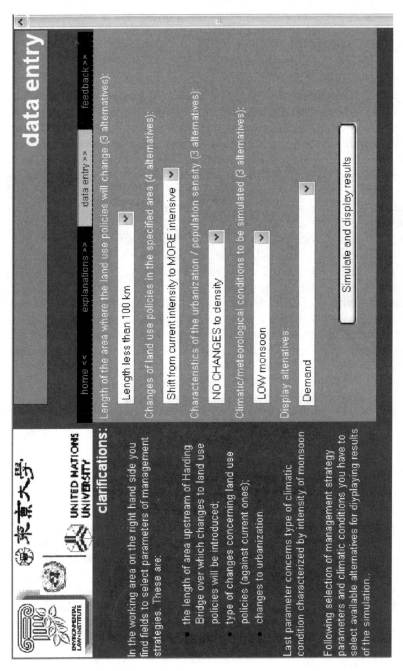

Figure 8.4 Example screen showing the selection of parameters for the strategic policy alternative.

average meteorological conditions, better than average conditions (more rainfall), and, finally, worse than average conditions (less rainfall).

The consequences of the selected policy alternative are represented by a monthly time series of the following indicators:
- normal water demand, which is the demand for water associated with current usage and unchanged conditions of land use in the area of interest (upstream of Harding Bridge);
- expected water demand, which is represented by the values of water demand calculated for the selected combination of decision variables;
- the normal water supply, equal to the flow rate at the Harding Bridge cross-section calculated for current (unchanged) land-use conditions; and
- the expected water supply, equal to the flow rate at the point of interest calculated for the user-selected land-use policy.

In addition, the user may select two other impact indicators, which are derived from values defined above, namely:
- the difference between the water supply and water demand calculated by the simulation model for unchanged land-use conditions; and
- the difference between the water supply and water demand calculated for selected land-use policy options.

Time series with all impact indicators are presented to the user in the form of a graph, which can also be printed out on a printer attached to the PC used to communicate with the DSS.

The system offers the user the ability to communicate with the developers of the DSS and provide them with feedback information. Feedback is provided in the form of a free text message which can be composed by a user and sent back. In order to obtain more specific feedback information from the DSS users, they are also asked to respond to a number of questions:
- the country they come from;
- their professional background and affiliation;
- their opinion about the information that should be presented in visual form; and
- their general opinion about the usability of the system.

Answers to these and eventually further (modified) questions will serve as the basis for improvements to the system and better understanding of the reactions of the public at large to tools such as this one. Thus, the materials and experiences collected within the framework of this study not only will allow this particular (prototype) system to be improved, but also will provide the basis for improvements in the design and implementation of similar tools to be developed for other case-study systems and for the formulation of future research agenda.

Summary

The experiences and lessons gained during the process of conceptualizing, designing and finally developing a prototype Web-based decision support system allow several conclusions to be formulated.

Environmental disputes and conflicts over the usage and sharing of natural resources can be solved in the framework of long and complex processes, where formal tools and models can contribute to the growth of mutual understanding and the objectification of the dispute by providing all parties with factual, accurate and verifiable information. A particularly important role can be played in this context by all efforts and developments involving the usage and popularization of Internet technology and Web-based tools and information sources. However, the development of Web-based tools is associated with challenges concerning two groups of factors: "soft" and "hard" ones.

The most important and the most challenging of the "soft" factors related to the human, organizational and political aspects of decision-making processes are as follows:
- the credibility of the models used in a DSS;
- acceptance of these models by all parties involved in solving the decision problem;
- the willingness of all parties involved in the dispute to communicate.

The "hard" challenges related to technical aspects of the DSS development are the following:
- the relevance of the models to the phenomena and processes under consideration;
- the availability of data;
- the speed of the data transfer;
- efficient handling of large amounts of data;
- the computational efficiency of the models and algorithms;
- the availability of computing power.

There is no simple strategy and approach to address all these challenges. However, thanks to continuous progress in the technological area, the "hard" challenges appear to be less difficult to resolve. The technology opportunities created by the growing popularity of the Internet and the availability of powerful computers and high-speed network technologies have led to the possibility of using distributed computers as a single, unified computing resource known as grid computing (see Baker et al. 2002). The term "grid" is chosen as an analogy to a power grid that provides consistent, pervasive, dependable and transparent access to electricity irrespective of its actual physical source. Computer grids enable the sharing, selection and aggregation of a wide variety of computing resources, including supercomputers, storage systems and data sources that are geographically scattered and owned by different organizations, for solving complex computational and data-intensive problems in science and engineering – problems related for instance to solving sophisticated decision problems associated with development planning in large, international river basins.

A computer grid can be viewed as an integrated, seamless computational environment providing its users with the following types of services:
- *computational services*, which provide secure services for executing applications on distributed computational resources individually or collectively;
- *data services*, which provide secure access to distributed databases and their management; to provide scalable storage and access to data sets, they may be replicated and catalogued; the processing of data sets is carried out using computational grid services, and such a combination is commonly called a data grid;
- *application services*, which are concerned with application management and providing transparent access to software and libraries;
- *information services*, which concern the extraction and presentation of data;
- *knowledge services*, which are concerned with how knowledge is acquired, used, retrieved, published and maintained to assist users and decision makers in achieving their particular goals and objectives – knowledge is understood in this context as information applied to achieve a goal, solve a problem or execute a decision.

Grid computing is currently at an early stage of development. It nevertheless offers enormous potential and the very necessary capabilities to provide a technical basis and infrastructure for resolving complex decisions relating to environmental and natural resources management problems. According to Huang and Chang (2003), there will continue to be attempts to apply new techniques and tools to environmental management. This tendency will definitely contribute to resolving the "hard" challenges associated with the development and creation of decision support systems in general, and in the Internet environment in particular. It would be worthwhile actively to explore this technological and research direction in a search for the powerful tools necessary to tackle complex decision problems related to water resources management.

The process of addressing and tackling the "soft" challenges mentioned above is more complex and not so straightforward, since progress in this area depends on so many different factors. Nevertheless, we can envision here an "ideal" scenario, in which a highly respected and unbiased international organization (for instance, the United Nations University) initiates, supports and coordinates international efforts aimed at developing and implementing a decision support system for selected river basins to assist the parties involved in a discussion (or even a dispute) over controversial strategic planning or development issues. Development of an appropriate DSS could help these parties to establish a sound dialogue and gradually resolve the controversies. This goal could be achieved in a step-by-step process involving modelling and technical efforts; building mutual understanding and trust between the parties involved in the dispute; establishing communication channels; and, finally, joint problem analysis and problem-solving using the capabilities offered by the DSS.

Acknowledgements

The research reported here was supported in part by the United Nations University, Tokyo University and the Environmental Law Institute. I would like to express my deep and sincere thanks to all those who helped to perform the research reported in this chapter. I am particularly grateful to Professor Mikiyasu Nakayama for encouraging me to undertake the research and for very deep and useful discussions during the writing of this chapter. Carl Bruch provided very helpful and stimulating motivation.

I owe special thanks to Neven Burazor for his programming support.

Note

1. Further information related to the subject of decision support systems can also be found on the Internet:
   - "Selected World Wide Web Sites For The Water Resources Professional", which contains numerous links to important water-related websites, at ⟨http://www.wrds.uwyo.edu/wrds/wwwsites.html⟩;
   - the "Water Resources Management" site of Delft University of Technology in the Netherlands at ⟨http://www.ct.tudelft.nl/wmg_land_water/⟩;
   - "An Inventory of Decision Support Systems for River Management" at ⟨http://www.geocities.com/rajesh_rajs/inventary.html⟩;
   - "The Environment Directory", which claims to be the world's biggest environment search engine, at ⟨http://www.webdirectory.com/⟩;
   - "Decision Support Systems Resources" at ⟨http://www.dssresources.com/⟩;
   - a list of water resources management and environmental models at ⟨http://www.wiz.uni-kassel.de/model_db/models.html⟩.

REFERENCES

Allan, R. B. and S. G. Bridgeman (1986) "Dynamic Programming in Hydropower Scheduling", *Journal of Water Resources Planning and Management* (ASCE) 112(3): 339–353.

Andreu, J., J. Capilla and E. Sanchis (1996) "AQUATOOL, A Generalized Decision-Support System for Water-Resources Planning and Operational Management", *Journal of Hydrology* 177: 269–291.

Baker, M., R. Buyya and D. Laforenza (2002) "Grid and Grid Technologies for Wide-Area Distributed Computing", *Software – Practice and Experience* 32(15): 1437–1466.

Biswas, A. K. and J. I. Uitto (2001) *Sustainable Development of the Ganges–Brahmaputra–Meghna Basins*. Tokyo: United Nations University Press.

Booch, G. (1994) *Object-Oriented Analysis and Design with Applications*, 2nd edn. Redwood City, CA: Benjamin/Cummings Publishing Company.

Chang, N. B. and Y. L. Wei (1999) "Strategic Planning of Recycling Drop-off Stations and Collection Network by Multi-Objective Programming", *Environmental Management* 24(3): 247–263.

Chang, N. B., H. Y. Lu and Y. L. Wei (1997) "GIS Technology for Vehicle Routing and Scheduling in Solid Waste Collection Systems", *Journal of Environmental Engineering* 123(9): 901–910.

Chen, J. C. and N. B. Chang (1998) "Water Pollution Control in the River Basin by Fuzzy Genetic Algorithm-Based Multi-Objective Programming Model", *Water Science and Technology* 37(8): 55–62.

Department of Civil Engineering, Colorado State University, and US Department of the Interior, Bureau of Reclamation, Pacific Northwest Region (2000) *MODSIM: Decision Support System for River Basin Management, Documentation and User Manual*. Fort Collins, CO, May.

Ellis, J. H. and C. S. ReVelle (1988) "A Separable Linear Algorithm for Hydropower Optimization", *Water Resources Bulletin* (AWRA) 24(2): 435–447.

Ford, D. T. and J. R. Killen (1995) "PC-Based Decision Support System for Trinity, Texas", *Journal of Water Resources Planning and Management* (ASCE) 121(5): 375–381.

Foster, I. (2002) "The Grid: A New Infrastructure for 21st Century Science", ⟨http://www.physicstoday.org/vol-55/iss-2/current.html⟩.

Fredericks, J. W., J. Labadie and M. Altenhofen (1998) "Decision Support System for Conjunctive Stream-Aquifer Management", *Journal of Water Resources Planning and Management*, March/April: 69–78.

Hipel, K. W., D. M. Kilgour, L. Fang and X. Peng (1997) "The Decision Support System GMCR in Environmental Conflict Management", *Applied Mathematics and Computation*, no. 83: 117–152.

Huang, G. H. and N. B. Chang (2003) "Perspectives of Environmental Informatics and Systems Analysis", *Journal of Environmental Informatics* 1(1): 1–6.

Ito, K., Z. X. Xu, K. Jinno, T. Kojiri and A. Kawamura (2001) "Decision Support System for Surface Water Planning in River Basins", *Journal of Water Resources Planning and Management* (ASCE) 127(4): 272–276.

Kazmi, A. and I. S. Hansen (1997) "Numerical Models in Water Quality Management: A Case Study for the Yamuna River (India)", *Water Science and Technology* 36: 193–197.

Keen, P. and M. Scott Morton (1978) *Decision Support Systems: An Organizational Perspective*. Reading, MA: Addison-Wesley Publishing.

Kelman, J., J. D. Damazio, J. L. Marien and J. P. Da Costa (1989) "The Determination of Flood Control Volumes in a Multireservoir System", *Water Resources Research* 25(3): 337–344.

Labadie, J. (1995) *River Basin Network Model for Water Rights Planning, MODSIM: Technical Manual*. Fort Collins, CO: Colorado State University, Department of Civil Engineering.

Li, S. and G. H. Chen (1994) "Modeling the Organic Removal and Oxygen Consumption by Biofilms in an Open-Channel Flow", *Water Science and Technology* 30: 53–62.

Loucks, D. P. and M. B. Bain (2002) "Interactive River-Aquifer Simulation and Stochastic Analyses for Predicting and Evaluating the Ecologic Impacts of Alternative Land and Water Management Policies", in M. V. Bolgov et al. (eds), *Hydrological Models for Environmental Management*, pp. 169–194. Dordrecht: Kluwer Academic Publishers.

Loucks, D. P. and K. A. Salewicz (1989) *IRIS – An Interactive River System Simulation Program, General Introduction and Description*. Laxenburg: International Institute for Applied Systems Analysis.

Loucks, D. P. and O. T. Sigvaldason (1982) "Multiple-reservoirs Operation in North America", in Z. Kaczmarek and J. Kindler (eds), *The Operation of Multiple Reservoir Systems*, IIASA Collaborative Proceedings Series CP-82-S3, pp. 1–104. Laxenburg: International Institute for Applied Systems Analysis.

Loucks, D. P., K. A. Salewicz and M. R. Taylor (1990) *IRIS – An Interactive River System Simulation Program, User's Manual Version 1.1*. Laxenburg: International Institute for Applied Systems Analysis.

Marino, M. A. and H. A. Loaiciga (1985) "Dynamic Model for Multireservoir Operation", *Water Resources Research* 21(5): 619–630.

Masliev, I. and L. Somlyody (1994) "Probabilistic Methods for Uncertainty Analysis and Parameter Estimation for Dissolved Oxygen Models", *Water Science and Technology* 30: 99–108.

Ministry of Land, Infrastructure and Transport, Infrastructure Development Institute (2001) *Study of Advanced Technology for Use of Global Geographic Information System, Creation of the Global Map Data Base for the Ganges River Basin and Trial Construction of the Simulation Models for Water Resources Management and Disaster Prevention Using Global Map Data*. Tokyo, Japan, March.

Minor, S. D. and T. L. Jacobs (1994) "Optimal Land Allocation for Solid and Hazardous Waste Landfill Siting", *Journal of Environmental Engineering* 120(5): 1095–1108.

Ning, S. K. and N. B. Chang (2002) "Multi-Objective, Decision-Based Assessment of a Water Quality Monitoring Network in a River System", *Journal of Environmental Monitoring* 4: 121–126.

Parker, B. J. and G. A. Al-Utaibi (1986) "Decision Support Systems: The Reality

That Seems Hard to Accept", *OMEGA International Journal of Management Sciences* 14(2): 135–143.

Sabherwal, R. and V. Grover (1989) "Computer Support for Strategic Decision-Making Processes: Review and Analysis", *Decision Sciences* 20: 54–76.

Salewicz, K. A. (2003) "Building the Bridge between Decision-Support Tools and Decision-Making", in M. Nakayama (ed.), *International Waters in Southern Africa*, pp. 114–135. Tokyo: United Nations University Press.

Salewicz, K. A. and D. P. Loucks (1989) "Interactive Simulation for Planning, Managing and Negotiating", in D. P. Loucks (ed.), *Closing the Gap between Theory and Practice*, IAHS Publication No. 180, pp. 263–268. Wallingford, Oxon: IAHS.

Shim, J. P., M. Warkentin, J. F. Courtney, D. J. Power, R. Shards and C. Carlsson (2002) "Past, Present and Future of Decision Support Technology", *Decision Support Systems* 33: 111–126.

Simonovic, S. P. (2000) "Tools for Water Management – One View of the Future", *Water International* 25(1): 76–88.

USACE [US Army Corps of Engineers] (1982) *HEC-5 Simulation of Flood Control and Conservation Systems, User's Manual*, CPD-5A. Davis, CA: Hydrologic Engineering Center.

――― (1985) *Reservoir System Analysis for Conservation HEC-3 User's Manual*, CPD-3A. Davis, CA: Hydrologic Engineering Center.

――― (1986) *HEC-5(Q) Simulation of Flood Control and Conservation Systems; Appendix, Water Quality Analysis*. Davis, CA: Hydrologic Engineering Center.

――― (1992) *Water Profile of Open or Artificial Channels HEC-2 User's Manual*, TD-26. Davis, CA: Hydrologic Engineering Center.

――― (1995) *HEC-RAS User's Manual*. Davis, CA: Hydrologic Engineering Center.

US Department of the Interior, Bureau of Reclamation, Pacific Northwest Region (2000) "River and Reservoir Operations Simulation of the Snake River – Application of MODSIM to the Snake River Basin", Report, May.

Venema, H. D. and E. J. Schiller (1995) "Water Resources Planning for the Senegal River Basin", *Water International* 20(2): 61–71.

Wang, M. and C. Zheng (1998) "Groundwater Management Optimization Using Genetic Algorithms and Simulated Annealing: Formulation and Comparison", *Journal of American Water Resources Association* 23(3): 519–530.

Zagona, E. A., T. J. Fulp, H. M. Goranflo and R. Shane (1998) "RiverWare: A General River and Reservoir Modelling Environment", in *Proceedings of the First Federal Interagency Hydrologic Modelling Conference*, Las Vegas, NV, pp. 5–113.

Zagona, E. A., T. J. Fulp, R. Shane, T. Magee and H. M. Goranflo (2001) "RiverWare", *Journal of American Water Resources Association* 37(4): 913–929.

Part III
Efforts by international organizations

9
Improving public involvement and governance for transboundary water systems: Process tools used by the Global Environment Facility

Alfred M. Duda and Juha I. Uitto

Background and focus

The World Commission on Water projected a gloomy future for freshwater basins and the people living in them, with an extra US$100 billion in investments needed each year in order to address basic water problems (WCW 2000). Conflicts between competing sectoral uses of water are becoming more common and are threatening internal as well as external security for many nations (Uitto and Duda 2002). The mismanagement has resulted in unprecedented degradation of ecosystems that nations depend on for economic and human security. River pollution and flow depletion have worsened to the point that they cross national borders and reach downstream coastal zones where poor communities suffer even further from a lack of access to clean water and natural resource depletion that limits their livelihoods. Nature does not neatly segment environmental or water problems by administrative boundaries or political units – 60 per cent of the water in our planet's rivers, half the Earth's land area and 43 per cent of its population are located in 261 transboundary freshwater basins (Wolf et al. 1999). Most of the large rivers of the world cross national borders, often resulting in water-use conflicts and tensions, as well as missed opportunities for sustainable development, peace and security.

Marine ecosystems also face critical problems and most of them are also transboundary in nature by virtue of interconnected currents or movement of living resources; 95 per cent of the global fisheries catch comes from 62 "large marine ecosystems" (LMEs) that parallel the continental shelves and potentially represent multi-country, ecosystem-based management units for reversing fisheries depletion (Duda and Sherman 2002). Not only has pollution from sewage, mud and nitrogen degraded marine ecosystems, but the conversion of coastal wetlands such as mangroves to short-lived aquaculture facilities has made the adverse impacts worse, as was noted by a United Nations oceans assessment (GESAMP 2001). Additionally, massive over-fishing of marine ecosystems with modern technology of distant factory fishing fleets has resulted in collapse around the world. At least 75 per cent of ocean fisheries are depleted, over-fished or fished at their limits (FAO 2000). Continued over-fishing in the face of scientific warnings, fishing down food webs, destruction of habitat and accelerated pollution loading have resulted in the dramatic collapse of coastal and marine ecosystems of both rich and poor nations (Pauly et al. 1998). Jackson et al. (2001) noted that ecological extinction caused by historical over-fishing has been more important than other causes of marine biomass and biodiversity depletion around the world, with existing populations being only a fraction of historical levels.

The key question is how to engage countries in effectively addressing environmental and water management concerns that cross national boundaries in an integrated manner (Duda and El-Ashry 2000; Duda and La Roche 1997).

This chapter focuses on the tools and processes used by the Global Environment Facility (GEF) in promoting improved public participation and better governance in managing transboundary water resources in the developing countries and countries with economies in transition. The tools and processes described include the use of national inter-ministerial committees; science-based transboundary diagnostic analyses; the development of politically acceptable strategic action programmes; public involvement; monitoring and evaluation (M&E); and local demonstration projects. Emphasis is placed on transparency in information-sharing and learning across initiatives around the world.

We draw upon the experiences gained during the first decade of the Global Environment Facility in promoting environmental protection and sustainable development around international waters and their drainage basins. Many of the conclusions are drawn from evaluations and studies carried out under the Facility's auspices (Bewers and Uitto 2001; Mee et al. 2004; Ollila et al. 2000). The authors argue that, if addressed correctly, the development of transboundary water resources will provide opportunities for cooperation between the countries sharing the resource. These cooperative opportunities have the potential of alleviating the tensions and potential conflicts between the riparians. In order to achieve these benefits, the participation of broad stakeholder groups and the public is a must.

Transparency and participation challenges

Instead of looking at multi-country freshwater and marine ecosystems as catalysts for cooperation toward sustainable development or for pursuing joint development options that collectively benefit all participating countries, some nations remain wary of basin-specific collaboration and consequently opportunities to secure the future. Fragmentation among institutions and disciplines, lack of cooperation among nations sharing transboundary ecosystems, as well as weak national policies, legislation and enforcement, all contribute to this challenge, involving the absence of joint governance.

Lack of political will, donor preferences for simple projects, lack of institutional transparency and limited finance have all limited progress. Excessive subsidies, incapable institutions, corruption and scientific uncertainties also contribute, as central governments, as a result of decentralization policies, have shrugged off responsibilities for reforms to lower levels of government without the capacity to take them on.

Similarly, inadequate or non-transparent information leads to wrong decisions. Watson and Pauly (2001) have reported recalculations of ocean fish catches that show a precipitous global decline since the 1980s. The authors identified inaccurate reporting of data to the Food and Agriculture Organization of the United Nations (FAO) that has likely distorted global assessments and subsequent policy. The inaccurate data have maintained a false sense of security through the years as burgeoning aquaculture replaced capture fisheries. Total production numbers have lulled policy makers into false impressions about the deepening global decline. Lack of transparency of information can be an important factor.

Demographic and health issues, especially concerns over food security, poverty and access to water and sanitation, constitute realities that many water-dependent communities face on a daily basis. Hardest hit by watershed and wetlands degradation, for example, are farmers and fishing groups who have to deal with depleted soils and declining fish catch resulting from sedimentation. It is a challenge to involve these groups in national programmes and seems impossible to undertake in transboundary situations.

Fragmented, thematic, single-purpose agency programmes are just not able to harness stakeholders sufficiently to drive reforms and gain support for important investment priorities. Country commitments to joint management regimes may actually create an "enabling environment" for the political driving forces needed for reforms. In reality, other priorities take centre-stage, integrated management is difficult to undertake, and progress seems slow, with fragmented approaches, donor-driven objectives and little funding to support time-consuming processes for reforms with few incentives.

The Global Environment Facility and international waters

The only new funding source to emerge from the 1992 Earth Summit, the Global Environment Facility (GEF) today counts 176 countries as members. Following a three-year pilot phase (1991–1994), the GEF was formally launched to forge cooperation and finance actions in the context of sustainable development that address critical threats to the global environment, such as biodiversity loss, climate change, degradation of international waters and ozone depletion. In 2002, land degradation and persistent organic pollutants (POPs) were added as new focal areas of the GEF.

Since its creation, the GEF has allocated more than US$4 billion in grants and leveraged an additional US$13 billion in co-financing from other sources to support more than 1,200 projects in over 140 developing nations and countries with economies in transition. In addition, the GEF has made more than 3,000 small grants (up to US$50,000 each) directly to non-governmental organizations (NGOs) and community organizations. GEF projects are implemented through a partnership among the United Nations Development Programme (UNDP), the United Nations Environment Programme (UNEP) and the World Bank. The Facility's policies are set by a Council representing all member nations.

Priorities for funding were established by the GEF Council in 1995 in its Operational Strategy (GEF 1996a). In the international waters focal area, the GEF aims to assist nations in resolving and preventing transboundary surface or groundwater problems as well as promoting the sustainable use of marine ecosystems by: (1) learning to work together; (2) identifying and adopting policy, legal and institutional reforms in the different economic sectors causing the degradation or use conflicts; and (3) testing the feasibility and effectiveness of on-the-ground priority investments to reverse transboundary degradation. Since 1991, 79 transboundary water projects have been funded with 135 different cooperating countries at a total cost of US$1.7 billion, including US$625 million in GEF grants. With this level of funding, the GEF is a very significant funding source for transboundary systems and is rapidly growing (Figure 9.1).

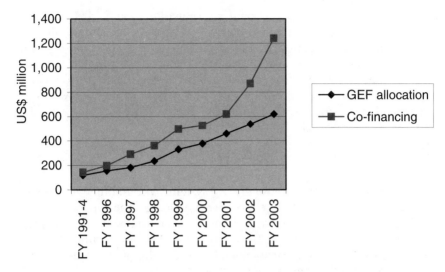

Figure 9.1 Growth of the GEF international waters portfolio since its establishment.
Source: GEF financial statistics.

The GEF Operational Strategy recommends that nations begin to address concerns about transboundary water systems by jointly undertaking strategic processes for analysing factual scientific information on transboundary issues, setting priorities and then determining the policy, legal and institutional reforms and investments needed to address the priorities in a country-driven strategic action programme (SAP). This facilitates a factual basis for supporting policy-making and fosters a place-based setting in which an ecosystem-based approach to management can be developed in conjunction with those with a stake in the outcome. These processes can be used to engage stakeholders within the area so that they contribute to the dialogue, become empowered to participate and influence policy debates.

Key tools and processes for participation

The GEF Operational Strategy recommends five key tools to be utilized by nations interested in addressing threats to their shared freshwater basins or LMEs: (a) national inter-ministerial committees for the GEF project; (b) production of a joint multi-country transboundary diagnostic analysis (TDA); (c) joint multi-country formulation of an SAP; (d) undertaking public involvement consistent with GEF policies; and (e) the establishment and utilization of M&E indicators to track progress and serve as a framework for adaptive management. A sixth tool, the use of local demonstrations developed as a result of lessons learned from early projects, helps to engage stakeholders and local communities with on-the-ground interventions. The following sections broadly describe the tools, with examples given in the case studies.

National inter-ministerial committees

Transboundary systems are usually big systems with multiple stresses in different sectors and different countries. Although the environment is the GEF's focus, many ministries other than environment have jurisdiction for needed interventions. Beginning with project preparation, inter-ministerial committees are recommended to develop a national-level coordination and collaboration capacity.

Transboundary diagnostic analysis

Lack of trust and empathy among nations can be overcome only by joint fact-finding and sharing of information so that all parties understand their mutual interconnectedness and have confidence in their partners. The GEF recommends beginning to address transboundary issues by jointly undertaking strategic processes for analysing factual, scientific information on transboundary concerns and their root causes in order to set priorities for action. This process has been referred to as a transboundary diagnostic analysis and it provides a handy tool for fostering participation at all levels, especially within the science community, which may be more progressive, may have access to technical information of relevance to decision-making, and may be able to stimulate transparency of information on transboundary issues.

Strategic action programmes

Once the issues and root causes are understood, countries are asked to identify national and regional policy, legal and institutional reforms and investments needed to address the priorities in developing a country-driven SAP. The process allows sound science to become the basis for policy-making and fosters a geographical location where an ecosystem-based approach to management can be developed. More importantly, the process can be used to engage stakeholders within the geographical area, including local communities and NGOs, so that they contribute to the dialogue and are empowered to participate. This process fosters cross-sectoral integration so that a truly ecosystem-based approach to improving management institutions may be pursued.

The national inter-ministerial committees should play a large role in the process of producing the SAP. This facilitates the development of country-driven, politically agreed ways ahead for commitments to action that address the priorities in a framework that encourages adaptive management. This shared commitment and vision for action has proven essential in GEF projects that have completed the processes in securing commitments for policy, legal and institutional reforms in different economic sectors. The GEF, together with other partners, may then fund implementation projects to assist countries in addressing the country-driven priorities for reform and investments as an incentive.

Public involvement policy

The need for public involvement – information dissemination, consultation and stakeholder participation – is set forth explicitly in the Instrument for the Establishment of the Restructured Global Environment Facility. The basic provisions of the Instrument state that all GEF-financed projects will "provide for full disclosure of non-confidential information, and consultation with, and participation as appropriate of, major groups and local communities through the project cycle".

It is seen that effective public participation is critical to the success of GEF-financed projects. When done appropriately, public involvement improves the performance and impact of projects by (GEF 1996b):
- enhancing recipient-country ownership of, and accountability for, project outcomes;
- addressing the social and economic needs of affected people;
- building partnerships among project-executing agencies and stakeholders; and
- making use of the skills, experiences and knowledge, in particular, of non-governmental organizations, community and local groups, and the private sector in the design and implementation and the evaluation of project activities.

Public involvement in the GEF context consists of three related and often overlapping processes: information dissemination, consultation and stakeholder participation. Stakeholders are defined as the individuals, groups or institutions that have an interest in the outcome of a GEF-financed project or those who are potentially affected by it. Stakeholder participation is defined to include the collaborative engagement of the various stakeholder groups in the identification of project concepts and objectives, the selection of sites, the design and implementation of activities, and the monitoring and evaluation of projects. Stakeholder participation is thus seen as essential throughout the project cycle. The GEF policy further notes the importance of paying attention to the needs of disadvantaged groups in and around project sites, e.g. indigenous communities, women and poor households.

It is important to provide relevant, timely and accessible information to as many stakeholders and stakeholder groups as possible. There is a need for both broad and project-specific consultations, especially at the local and subnational levels. Awareness-raising and capacity development amongst the various stakeholder groups are seen as essential. According to GEF policies, all public involvement activities must be conducted in a transparent and open manner.

Monitoring and evaluation

The M&E framework adopted by the GEF is based on the realization that environmental improvement in the international waters setting is a long-term process. The actual results of actions taken today will often take years or even decades to materialize. This mandates the establishment of interim indicators that allow the measurement of whether the interventions are moving to the desired direction.

The agreed international waters M&E framework utilizes three levels of indicators (Duda 2002):
- process indicators
- stress reduction indicators
- environmental status indicators.

Process indicators track the implementation of the agreed processes amongst the countries and stakeholders. These processes are usually concerned with political processes and the policy, legal, regulatory and institutional reforms that are needed to address the environmental problems facing the water body. Process indicators are often used to track progress in the implementation of the TDA and SAP. The processes to be tracked may take place either at the multi-country or at the individual country level. In some cases, the key process requires actions by one country within the shared water body's drainage basin. An example would be the promulgation of a law to regulate activities that pollute the water body in question. In others, it is necessary for all of the riparian countries to act on the process, such as harmonizing their legislation. Nevertheless, even in this latter case the required action is to be taken by individual countries within their own legislations.

Some processes will by definition require joint action by the countries around the water body. An example would be the ratification of an SAP or specific agreements that will help its implementation. Process indicators thus track that the riparian countries adhere to the agreements with regard to the establishment of policies, institutions and legal frameworks that will be conducive to sustainable development of the shared water body.

Stress reduction indicators go a step further in focusing on the actual implementation of measures that will reduce the environmental stress to the water body. These actions often fall within the purview of individual countries, such as the installation and operation of a sewage treatment system for an industrial plant or an urban centre.

Once these investments are operational, the actual discharges to the water body will be reduced, eventually leading to reduced environmental stress and improved status of the ecosystem. Environmental status indicators measure these actual environmental improvements. Because achieving the actual environmental improvements usually takes a long time, most of the impacts that can be expected during a project's lifetime are at the process and stress reduction levels.

M&E can be a powerful tool for tracking the performance of the riparian countries and various stakeholders in implementing the processes and stress reduction measures that are needed to improve the environmental status of a shared water body (Uitto 2004). Indicators can be used as objective measures to track compliance with and implementation of multi-country agreements. They can also be used to identify those sectors and actors responsible for non-compliance. It is therefore important for the various stakeholders to agree on the desired outcomes, targets and indicators. These can then form the basis against which performance is measured. A transparent system that provides tracking information on progress to the various groups of stakeholders can effectively promote public participation and improved governance.

Local demonstration projects

Long-term environmental improvements may not always be the immediate priority of local people concerned with improving their material standard of living. Therefore, there is always a risk that projects dealing with global environmental objectives are sidelined or even conflict with local goals. As noted in Ollila et al. (2000), a number of GEF international waters projects supported local demonstration activities that captured the interest of local people and made their leaders and country officials comfortable about recommending reforms and investments that would replicate the demonstrations. These demonstration projects stimulated local interest and participation in the projects.

Freshwater basin case studies

Through their GEF participation, countries are learning that transboundary water management is about much more than just sharing water. They are learning that their shared drainage basin is the key element, with its environmental assets, its communities and its land resources. Land-use decisions are often in reality water-use decisions, and the security of downstream communities and downstream economies can be placed in jeopardy because of misuse of the land or water. Sustainable development of transboundary basins is all about sharing the benefits from improved land and water resources management in basins, not just about dividing up limited amounts of water.

Examples from Latin America: The Bermejo and San Juan basins

A good example of reforms and investments involving both land and water resources exists in the GEF-funded and UNEP-implemented Bermejo Binational Basin project in Argentina and Bolivia. The basin suffers from droughts and floods, and strategic planning processes combined with on-the-ground demonstrations of water harvesting and soil erosion control empowered communities to participate in determining their sustainable future. With access to resources being the main problem for poor communities, institutions that affect their lives must also be accessible to them. These are not only local but national and basin-wide as well, so that enabling conditions are created for long-term local participation. In the Bermejo project, local NGOs were involved in planning activities and were consulted in processes that provided access to the commissions and steering committees to influence programmes. The recent external evaluation of the GEF international waters focal area also found that the project had convinced the government of Argentina to work at the decentralized level in order to achieve the full benefits at that decentralized level (Mee et al. 2004).

An initial two-year project supported development of the TDA/SAP, coupled with demonstration activities in basin management and land degradation control. A key objective was to involve local stakeholder groups in the basin in determining their sustainable development future.

The project aroused considerable involvement and excitement among NGOs and subnational levels of government as well as the binational commission for the basin, with over 2,000 people participating (UNEP 2000). This lesson about the participation of local stakeholders in the identification and planning of both demonstration activities as well as the necessary multi-country strategic work (TDA and SAP) is important for the commitment to implement integrated approaches to land and water management. The attention of local community stakeholders was achieved by pilot interventions in improved agricultural practices, water harvesting, pasture and grazing improvement, and reforestation for watershed stabilization and carbon sequestration. These made the basin's land and water management problems and the potential solutions more concrete to the wider public, including farmers, whose poor land management practices created the transboundary sedimentation problems to begin with.

For example, in the Tolomosa watershed, over 100 dikes and water-harvesting devices were built to reduce sediment transport to the San Jacinto reservoir. The structures harvested and trapped rainwater and created oases in the semi-arid landscape, supporting the irrigation of nurseries and the watering of livestock for local livelihoods. The pilots attracted the interest of other stakeholders, who have replicated the successes and thus served as teaching/learning tools for local communities. These successful demonstrations may reduce the risk of implementation failure by creating community buy-in and on-the-ground development benefits.

Following completion of the TDA/SAP in 2000, a follow-on implementation project was approved by the GEF Council in 2001 for US$20 million (US$11 million GEF) to assist with implementing priority measures identified by the SAP. The project broadens and deepens the interventions in the basin, addressing land degradation, biodiversity and the fluctuating climate in a collective manner, including a protected transboundary corridor as requested by the local communities.

Another example in Latin America is the San Juan River basin, where UNEP and the Organization of American States are assisting Costa Rica and Nicaragua with an initial project featuring all six tools mentioned above. Even more enthusiastic participation was found in this basin, despite boundary disputes between the countries. Universities are engaged on both sides of the border and more than 250 institutions – from farmer cooperatives to schools, local municipalities and NGOs – were involved in the execution of specific project components to ensure active participation. Various legal-administrative instruments had to be signed for the activities, creating an official status for activities, including a mandate for gender-specific activities important for rural development. In addition to various outreach workshops, a periodic newsletter *Procuenca-San Juan Hoy* is published by local groups to engage the populace. A website has proved to be an important way for the universities and different ministries to share information.[1] The multi-stakeholder participation has promoted principles of openness and inclusiveness. Direct participation of key stakeholders at all levels of society has fostered the principle of responsibility because different groups were accountable for different activities. Central and local governments have shared information and results with all stakeholders so that the project work has been transparent and government was held accountable.

Examples from Europe: The Danube–Black Sea basin

The Black Sea LME has 17 countries that drain to it, 15 of them eligible for GEF assistance. Germany and Austria are also located in the basin and contribute significantly to the key transboundary pollution problems plaguing the Danube delta and Black Sea – coastal nutrient overenrichment or eutrophication, especially nitrogen (Mee 1992). A number of other key transboundary issues, such as hotspots of toxic substances and over-fishing, were identified in both the Danube and the Black Sea basin TDA and SAP.

Initial GEF assistance to the region began in the early 1990s when two sets of nations (13 for the Danube basin and 6 for the Black Sea) began joint planning activities under recently signed regional environmental conventions. The Danube basin effort was initially separate from that for the Black Sea, and resulted in a plan that did not sufficiently address the downstream pollution impacts on the Black Sea. A small project was then approved by the GEF to undertake the TDA/SAP process following adoption of the GEF Operational Strategy in 1995. The processes were successful in the upstream Danube basin countries, which agreed to make commitments on policy, legal and institutional reforms and investments in the agricultural, municipal, industrial and environmental sectors in their SAP. The focus is on nitrogen reduction from land-based sources in order to restore the Danube delta and the Black Sea. Water-related needs for reducing pollution from land management activities are given priority in reform programmes and investments.

Of great interest here is the role played by NGOs in the Danube Environmental Forum and the potential for participation represented by a new legal mechanism: the Convention on Access to Information, Public Participation in Decision-making and Access to Justice in Environmental Matters, adopted on 25 June 1998 in the Danish city of Aarhus at the fourth Ministerial Conference in the "Environment for Europe" process under the auspices of the United Nations Economic Commission for Europe (UNECE).[2]

The Aarhus Convention is considered to be a new kind of environmental agreement that links environmental and human rights. It is based on the notion that we owe an obligation to future generations. It also emphasizes that sustainable development can be achieved only through the involvement of all stakeholders. The Convention links government accountability, transparency and responsiveness to environmental protection. The Convention grants rights to the public, and imposes on the parties and public authorities obligations regarding access to information and public participation and access to justice.

As a practical application, the Protocol on Pollutant Release and Transfer Registers was adopted at an extraordinary meeting of the Parties to the Convention on 21 May 2003. The meeting took place within the fifth "Environment for Europe" Ministerial Conference in Kiev, Ukraine. The Protocol is the first legally binding international instrument on pollutant release and transfer registers. Its objective is stated as: "to enhance public access to information through the establishment of coherent, nationwide pollutant release and transfer registers (PRTRs)". The PRTRs are inventories of pollution from industrial sites and other sources. Although the Protocol is concerned with information transparency, rather than regulating pollution per se, it is anticipated that it will have a significant effect on pollution control by putting public pressure on companies that are identified as the biggest polluters.

The Protocol requires each party to establish a mandatory PRTR based on annual reporting that:
- is publicly accessible through the Internet, free of charge;
- is searchable according to separate parameters (facility, pollutant, location, medium, etc.);
- is user-friendly in its structure and provides links to other relevant registers;
- presents standardized, timely data in a structured, computerized database;
- covers releases and transfers of at least 86 pollutants covered by the Protocol, such as greenhouses gases, acid rain pollutants, ozone-depleting substances, heavy metals, and certain carcinogens, such as dioxins;
- covers releases and transfers from certain types of major point sources (e.g. thermal power stations, mining and metallurgical industries, chemical plants, waste and waste-water treatment plants, paper and timber industries);
- accommodates available data on releases from diffuse sources (e.g. transport and agriculture);
- has limited confidentiality provisions; and
- allows for public participation in its development and modification.

Just like the Convention, the Protocol sets minimum requirements, meaning that parties can include additional pollutants and facilities, and the parties to the Protocol are required to work towards convergence between PRTR systems. All states can sign and ratify the Protocol, including those that have not ratified the Convention and those that are not members of the UNECE. Consequently, it could develop into a global protocol. So far, 36 countries and the European Communities have signed the Protocol. There have not yet been any ratifications.

Example from Africa: The Lake Tanganyika basin

Lake Tanganyika is the fourth-largest lake in the world. The UNDP assisted Burundi, the Democratic Republic of Congo (DRC), Tanzania and Zambia in addressing transboundary degradation of their shared lake basin through a GEF project in the late 1990s. High-level officials from each nation participated in a Steering Committee responsible for the project. Various programmes were established with the objective of helping the riparian countries produce an effective and sustainable system for managing and conserving the biodiversity of the lake. By involving local communities in its design, the programmes embraced the dual needs of development and conservation so that people's livelihoods could be maintained into the future. The programmes varied from biodiversity to fisheries, the impacts of sedimentation, catchment degradation, pollution, economic issues, education and development of a joint geographical information system (GIS) for sharing scientific and management data.

The international waters programme evaluation commended the Lake Tanganyika project, suggesting that its success was attributable to the high level of ownership at all levels. This enabled it to overcome very difficult conditions caused by armed conflict, the high prevalence of HIV-AIDS and widespread poverty (Mee et al. 2004).

The project adopted the approach of joint fact-finding in compiling information so that all countries could review it and update it through GIS technology. The resulting TDA sets priorities for two or three top-priority shared water issues based on existing science. Pollution discharges in Bujumbura, Burundi, and in Kigoma, Tanzania, were cited as hotspots for abatement activities. Excessive sediment loading from certain river basins, mostly in Burundi, the DRC and Tanzania, was determined to be a priority for accelerated attention, and over-fishing was identified as important because of the large commercial fishery, its economic importance to certain nations, and the transboundary nature of the stock and pattern of landings and markets.

The project formulated an SAP that addressed integrated land and water resources management in the basin to reduce the effect of eroded soils on the lake and to reduce stress through fisheries reforms. Publications are available on the project's website,[3] which, because of the project's distance from national capitals, has been an essential communications and participation tool. A firewall was established for internal use, which was useful for exchanging information between the countries in this remote area. Of note has been the broad network in the scientific community within the countries and abroad that has been involved with the project and brought the best available science to management decision-making.

The Lake Tanganyika governments completed the fourth draft of an international treaty to affirm their political support for the restoration and protection of the Lake Tanganyika ecosystem by the end of the first project. During GEF-funded preparation of a new project to implement the SAP, they completed the negotiations and in 2003 signed the Convention on the Sustainable Management of Lake Tanganyika. The Convention establishes a Lake Tanganyika Authority consisting of a joint Management Committee and a Secretariat to assist the nations in achieving sustainable management of the lake basin, in conserving its biological resources, and in reversing degradation of the catchment area draining to the lake so that it can cope with the fluctuating climate. Various protocols specify progressively more stringent country commitments as implementation proceeds. A GEF project to implement this collective approach to land and water resources management is under development.

Marine ecosystem case studies

Across Africa, Asia and the Pacific, Latin America and the Caribbean, and in Eastern Europe, country officials have been experimenting with the GEF to reverse the decline of their marine ecosystems, testing methods for restoring once abundant biomass in order to sustain growing populations of coastal communities, and to conserve highly fluctuating systems to ensure continued benefits for future generations.

The geographical area of an LME, its coastal area and contributing basins would constitute the geographical area for assisting states to understand linkages among root causes of degradation and then integrating the necessary changes into sectoral economic activities. These LME areas serve as a platform to begin capacity-building and for making pragmatic use of science in improving the management of coastal and marine ecosystems. LMEs represent a scale of effort that can complement projects at other scales, such as local integrated coastal management (ICM) initiatives and drainage basin programmes.

The South China Sea and East Asian seas

The UNEP-implemented project on Reversing Environmental Degradation Trends in the South China Sea and Gulf of Thailand is concerned with creating an environment at the regional level that fosters and encourages collaboration and partnership, between all stakeholders and at all levels, in addressing the environmental problems of the South China Sea and that enhances the capacity of the participating governments to integrate environmental considerations into national development planning (UNEP/GEF 2003). A recent review of the project found that stakeholder participation in the project had been highly satisfactory from the regional level to the local level (Harstad et al. 2004). Strong country ownership was secured through mechanisms that include inter-ministerial committees, national technical working groups and the systematic involvement of local and subnational authorities and non-governmental stakeholders at the demonstration sites.

The project was programmed in conjunction with two other GEF international waters projects to fit in with the attempt to restore and protect the globally significant coral reefs, sea grass beds, mangroves and wetlands of the LME and its coast (EAS/RCU 2000). The Mekong Basin project, with its valuable delta, receives GEF assistance through the World Bank, and the hotspot ICM demonstration activities conducted through a programme entitled Building Partnerships for the Environmental Protection and Management of the East Asian Seas (PEMSEA) are also an integral part of the GEF's programmatic approach for that highly threatened LME. Whereas the South China Sea project undertakes collective strategic processes for developing a more ecosystem-based approach to management through the production of a TDA and an SAP, PEMSEA has supported a number of complementary local demonstrations of ICM since 1996 that are well known throughout the ICM community (Chua 1998). The South China Sea project illustrates the application of M&E indicators that were established in advance to track progress in the restoration of coastal ecosystems and future protection for them.

In the East Asian seas, one of the world's major centres of marine biodiversity, several major rivers discharge into semi-enclosed seas. Pressures from urbanization, agricultural runoff and population growth continue to grow in its coastal areas, resulting in habitat destruction and locally severe nutrient, sewage and industrial contamination in such areas as the Bohai Sea, the Gulf of Thailand and Manila Bay. The lack of regional agreements for controlling discharges and for managing the marine environment is attributed to territorial disputes as well as to insufficient coordination of policies and programmes.

One solution introduced by the GEF-financed project PEMSEA, and implemented by the UNDP, is to make use of integrated coastal zone management (ICZM) to formulate region-wide enabling policies and programmes and capacity-building. This ICZM approach is supplemented at the local level by two demonstration sub-projects in the Philippines and China. By field-testing community-based approaches for water pollution control, beach cleanup and sustainable fisheries, the project was able to translate lessons in the field into national and regional policies and programmes. One of the results is the realization of the importance of providing incentives for private businesses to invest in pollution control and cleanup facilities. The experience in China also showed the value of community-based administrative guidelines for access to water and fisheries. China piloted the zoning of sea-space to resolve and prevent conflicts in coastal area use among stakeholders. The successful pilot has now been incorporated into China's Marine Law and will empower stakeholder involvement in sea-space zoning processes for its entire coast (Chua 1998). The project has thus directly contributed to the decentralization of marine and coastal management to local governments (Chen and Uitto 2003).

PEMSEA assisted the government of the Philippines as it developed the Manila Bay Declaration and the Manila Bay Coastal Strategy for its part of the shared South China Sea. This complementary initiative is multi-jurisdictional in nature, involving the respective national governments, provinces in the drainage area and the large municipalities of Manila, and it is an equivalent of an SAP under GPA (Global Programme of Action for the Protection of the Marine Environment from Land-based Activities) for the contributing freshwater basin that is enacted in the framework of coastal sustainable development. The political declarations have been adopted at the highest level and represent a decade-long commitment to action with stakeholders of the area. These processes have ensured transparency and the opportunity for participation.

The Pacific small island developing states

Even in the rich tuna fisheries of the western Pacific, UNDP (1998) reported that Pacific small island developing states (SIDS) received only about 4 per cent of the value of the tuna taken by distant fleets. The western Pacific marine ecosystem is nevertheless the lifeblood of the Pacific SIDS economies. Heads of state of the 13 Pacific SIDS adopted their GEF SAP in September 1997 and began implementation of their GEF/ UNDP international waters project thereafter. Although a number of components were involved – including interventions addressing community water supplies and waste management, integrated watershed management and marine protected areas – a main part of the project supported the countries through the Pacific Forum Fisheries Agency to negotiate a regional convention on the conservation, management and sustainable use of their highly migratory fish stocks. A commission is being established to oversee a more ecosystem-based approach to management.

Known as the Convention on the Conservation and Management of Highly Migratory Fish Stocks of the Western and Central Pacific Ocean, the Convention is a model of its kind. The GEF assistance helped level the playing field among the Pacific SIDS as they negotiated the Convention with Asian, North American and European nations. Following seven sessions of what was known as the MHLC (Multilateral High Level Conferences on South Pacific Tuna Fisheries) process (Sydnes 2001), the Convention was signed in September 2000 and was the first agreement to be successfully negotiated on the basis of the 1995 UN Fish Stocks Agreement and consistent with the United Nations Convention on the Law of the Sea (UNCLOS).

The importance of learning among projects

The GEF has found that South–South exchanges of experience and structured learning among the various transboundary waters projects are quite valuable, in that the projects learn from each other, especially with regard to managed strategies such as stakeholder involvement and processes for stimulating participation. Such learning has the potential to break down barriers to trying such activities in projects as experiences are shared, resources for capacity-building are found, and confidence increases to implement involvement strategies. A UNDP-implemented project known as IW:LEARN has tested the latest web-based technology to assist GEF transboundary waters projects in exchanging experiences and transferring knowledge of the technology.

IW:LEARN aims to build a "global knowledge community" among GEF projects to sustain the Earth's transboundary water resources. Specific services provided to foster this international waters community of practice include:
1. facilitated face-to-face and electronic forums among international waters managers and among stakeholders to identify and address priority transboundary waters management needs at the local, national, regional and global scale;
2. synthesis of "knowledge products" (e.g. articles, guidelines, distance education modules) gleaned from instructive experiences and lessons learned in order to address to these needs;
3. dissemination of these knowledge products via both on-line and off-line electronic media as well as through face-to-face workshops and outreach activities;
4. development of on-line and standalone electronic "resource centres"

to provide wide access to these knowledge products and related knowledge resources (e.g. international waters project profiles, tools, best practices, community news, events) via both electronic and traditional media (paper, radio, etc.);
5. collaboration with international waters projects to test and evaluate emerging information and communications technologies (ICTs) and processes to advance transboundary water management;
6. needs-based technical assistance to projects to apply such ICTs to increase the effectiveness of transboundary communication and coordination within and between projects;
7. workshops for international waters project personnel to develop and replicate all the above products, services and tools to meet their own transboundary waters management needs; and
8. establishment of regional support facilities to assist personnel in the development of these products and services to foster additional regional and thematic knowledge communities for the benefit of international waters projects in their region.

IW:LEARN has supported forums and dialogues among over 200 participants in international waters projects and their civil society counterparts at the global scale. An on-line resource centre has been deployed by IW:LEARN and its partners, the "International Waters Resource Centre".[4]

Lessons learned from GEF transboundary projects

This chapter has outlined the important tools and processes utilized by the GEF to ensure improved governance and stakeholder participation in its international waters operations. Use of these tools and processes has been highlighted through examples from several freshwater and marine international waters projects. Several lessons can be learned from these operations and have been confirmed by evaluations and studies carried out in recent years.

A key lesson is that it is essential to work on three levels simultaneously: regional, national and subnational. Only through site-specific, location-based river, aquifer or LME transboundary partnerships on joint resources management can the transition to collective, sustainable use of these large, multi-country water systems be achieved. Reaching regional consensus on the issues through joint science-based fact-finding is thus important. The TDA process also leads to good practices, such as prior notification of activities by one of the riparians affecting the joint environment; transparency; and dialogue with the scientific community and stakeholders in the riparian countries.

However, most of the actions to relieve stress on the transboundary water body need to be undertaken by individual countries. Because most environmental stresses are caused by productive sectors, it is crucial to involve all relevant ministries and sectoral agencies. In this task, the establishment of inter-ministerial committees for the implementation of the politically agreed SAP has proven to be a good way of ensuring buy-in from the productive sectors. To be effective, the SAP should set targets and environmental quality objectives that can be used to foster policy, legal and institutional reforms and good governance. These steps can then be tracked in a transparent and participatory manner.

At the subnational level, on-the-ground demonstrations bring about local benefits and allow learning to take place. They are irreplaceable in fostering local stakeholder buy-in and involvement.

The M&E requirements, including the establishment of indicators, contribute to transparency of information. They also promote capacity development and technology transfer and ensure that management institutions are engaged with the science community in joint efforts developed in conjunction with stakeholders. Similarly, the exchange of information and lessons between similar projects around the world can significantly contribute to improved planning and operations.

Conclusions

The growing number of country-driven commitments to change, as fostered by the GEF, and the global imperative to change because of the degraded condition of the world's coastal oceans and transboundary basins provide an unprecedented opportunity for accelerating the transition to the sustainable use, conservation and development of freshwater, coastal and marine ecosystems. The costs of inaction are much too high not to support the fledgling efforts of over 135 countries focusing on specific, shared LME and freshwater basins. Momentum must not be lost because the result may be irreversible damage to freshwater, coastal and marine ecosystems, the poor communities that depend on them and the economies of nations.

Regional, multi-country partnerships driven by the governments' desire to collaborate around public goods and sustainable development are essential in order to achieve coherence among donors and financial institutions in assisting groups of nations collectively to share the benefits from development in basins and marine ecosystems. Partnerships can simultaneously contribute domestic, regional and global benefits in reducing the disease burden of the poor resulting from unclean water and an unclean environment, securing their livelihoods for poverty reduction, and simultaneously resolving transboundary water concerns, restoring coastal biomass and biodiversity as global public goods and sustaining the natural resource capital upon which economies are based. The partnerships need to be developed for sequencing reforms, building capacity for the range of required programmes and reforms, and fostering investments that can help to balance conflicting basin uses and support the transition to sustainable development. Reallocation of phased reductions in agricultural subsidies and fisheries subsidies by 2015 and increased resources for conserving natural capital as a public good through the Highly Indebted Poor Countries Initiative and the World Bank's International Development Association would provide sufficient baseline finance for the partnerships. Reversing the degradation of transboundary freshwater systems and the depletion of coastal oceans is fundamental to sustainable development.

To accomplish this, public involvement, stakeholder participation and reforms in governance are essential first steps. The GEF projects illustrate that holistic, ecosystem-based approaches to managing human activities in LMEs, their coasts and transboundary basins are critical and need a place-based platform to focus stakeholder interest on the multiple benefits available under multiple global instruments. Instead of establishing competing programmes with their inefficiencies and duplication, which is currently the norm, GEF projects foster action on priority transboundary issues across instruments in an integrated manner. In fact, the adaptive management framework resulting from iterative application of the GEF Operational Strategy allows for sequential capacity development, stakeholder participation, technology introduction, governance reforms and investments, so that a collective response to global conventions and other instruments can be achieved in a practical manner.

The science-based approach, combined with strong institutional dimensions, may make it easier to handle ecological surprises in the future, such as those generated by a fluctuating climate, and may make it possible to insulate the poor communities that are the first to suffer the adverse effects of inadequate and inappropriate management efforts. Developing such location-based, collective responses to global driving forces for change would be impossible without the necessary governance institutions and the mobilization of the different levels of stakeholders needed for decision-making.

Notes

1. See ⟨http://www.oas.org/sanjuan/⟩.
2. See ⟨http://www.unece.org/env/pp/⟩.

3. "The Lake Tanganyika Biodiversity Project" at ⟨http://www.ltbp.org⟩.
4. See ⟨http://www.iwlearn.net⟩.

REFERENCES

Bewers, J. M. and J. I. Uitto (2001) *International Waters Program Study*. Evaluation Report No. 1-01. Washington, DC: Global Environment Facility.

Chen, S. and J. I. Uitto (2003) "Governing Marine and Coastal Environment in China: Building Local Government Capacity through International Cooperation", *China Environment Series*, 6, pp. 67–80. Woodrow Wilson International Center for Scholars, Environmental Change and Security Project.

Chua, T.-E. (1998) "Lessons Learned from Practicing Integrated Coastal Management in Southeast Asia", *Ambio* 27(8): 599–610.

Duda, A. M. (2002) *Monitoring and Evaluation Indicators for GEF International Waters Projects*. Monitoring and Evaluation Working Paper 10, Global Environment Facility, Washington, DC.

Duda, A. M. and M. T. El-Ashry (2000) "Addressing the Global Water and Environmental Crises through Integrated Approaches to the Management of Land, Water, and Ecological Resources", *Water International* (25): 115–126.

Duda, A. M. and D. La Roche (1997) "Joint Institutional Arrangements for Addressing Transboundary Water Resources Issues – Lessons for the GEF", *Natural Resources Forum* 21: 127–137.

Duda, A. M. and K. Sherman (2002) "A New Imperative for Improving Management of Large Marine Ecosystems", *Ocean and Coastal Management* 45: 797–833.

EAS/RCU [East Asian Seas Regional Coordinating Unit] (2000) *Transboundary Diagnostic Analysis (TDA) for the South China Sea*. EAS/RCU Technical Report Series No. 14.

FAO [Food and Agriculture Organization of the United Nations] (2000) *The State of the World Fisheries and Aquaculture*. Rome: FAO, Fisheries Department.

GEF [Global Environment Facility] (1996a) *Operational Strategy*. Washington, DC: GEF.

―――― (1996b) *Public Involvement in GEF-financed Projects*. Washington, DC: GEF.

GESAMP [IMO/FAO/UNESCO-IOC/WMO/WHO/IAEA/UN/UNEP Joint Group of Experts on the Scientific Aspects of Marine Environmental Protection and Advisory Committee on Protection of the Sea] (2001) *Protecting the Oceans from Land-Based Activities – Land-Based Sources and Activities Affecting the Quality and Uses of Marine, Coastal, and Associated Freshwater Environment*. Arendal, Reports and Studies No. 71.

Harstad, J., S. Graslund and J. I. Uitto (2004) *Report of the Review of the UNEP/GEF Project: Reversing Environmental Degradation Trends in the South China Sea and Gulf of Thailand*. GEF Specially Managed Project Review (SMPR). Washington, DC: Global Environment Facility Office of Monitoring and Evaluation.

Jackson, J. B. C. and 18 others (2001) "Historical Overfishing and the Recent Collapse of Coastal Ecosystems", *Science* 293: 629–638.
Mee, L. D. (1992) "The Black Sea in Crisis: A Need for Concerted International Action", *Ambio* 22: 278–286.
Mee, L., J. Okedi, T. Turner, P. Caballero and M. Bloxham (2004) *International Waters Program Study*. Washington, DC: Global Environment Facility Office of Monitoring and Evaluation.
Ollila, P., J. I. Uitto, C. Crepin and A. M. Duda (2000) *Multicountry Project Arrangements: Report of a Thematic Review*. Monitoring and Evaluation Working Paper 3. Washington, DC: Global Environment Facility.
Pauly, J. D., V. Christensen, J. Dalsgaard, R. Froese and F. Torres, Jr. (1998) "Fishing Down Marine Food Webs", *Science* 279: 860–863.
Sydnes, A. K. (2001) "Establishing a Regional Fisheries Management Organization for the Western and Central Pacific Tuna Fisheries", *Oceans and Coastal Management* 44: 787–811.
Uitto, J. I. (2004) "Multi-country Cooperation around Shared Waters: Role of Monitoring and Evaluation", *Global Environmental Change* no. 14: 5–14.
Uitto, J. I. and A. M. Duda (2002) "Management of Transboundary Water Resources: Lessons from International Cooperation for Conflict Prevention", *Geographical Journal* 168(4): 365–378.
UNDP [United Nations Development Programme] (1998) *Implementation of the Strategic Action Programme of the South Pacific Developing States*. Project Document UNDP/RAS/98/G32/A/1G/99. New York: UNDP.
UNEP [United Nations Environment Programme] (2000) *In-depth Evaluation of the UNEP/GEF Project GF/1100-97-07: A Strategic Action Programme for the Binational Basin of the Bermejo River*. Nairobi: UNEP.
UNEP/GEF (2003) *Project on Reversing Environmental Degradation Trends in the South China Sea and Gulf of Thailand*, ⟨http://www.unepscs.org⟩.
Watson, R. and D. Pauly (2001) "Systematic Distortions in World Fisheries Catch Trends", *Nature* 414: 534–536.
Wolf, A. T., J. A. Natharius, J. J. Danielson, B. S. Ward, J. K. Pender (1999) "International River Basins of the World", *Water Resources Development* 15(4): 387–427.
WCW [World Commission on Water] (2000) *A Water Secure World – Vision for Water, Life, and the Environment*. World Commission on Water Report, World Water Council.

10

Public participation and governance: A Mekong River basin perspective

Prachoom Chomchai

Introduction

The Mekong River, known in China as Lancangjiang, rises in Tibet (or Xizang, as it has been renamed by the Chinese) and flows south through China, whose southernmost province is Yunnan. It then relentlessly pursues its southerly course, serving first as the joint boundary between Myanmar (formerly Burma), Laos and Thailand at the infamous Golden Triangle and then as an occasional boundary between Thailand and Laos as well as a domestic watercourse in Laos before cutting across Cambodia and southern Viet Nam. Known in legal parlance as a successive as well as a contiguous river, the Mekong empties into the South China Sea after traversing its delta, which is shared by Cambodia and Viet Nam. With its average annual discharge of 500 billion m^3, its length of 4,800 km and a basin area of 795,000 km^2, the Mekong River constitutes one of South-East Asia's most substantial complexes of resources. When its potential is expressed in terms of energy equivalents, the river may be said to represent an oil-well turning out no fewer than 1.5 million barrels of crude per day, albeit with a difference: compared with the output of a typical oil-well, this essentially solar source of energy is non-polluting, inexpensive and fully renewable (provided, of course, that the productivity of its catchment area is maintained). The basin has a population of about 70 million (as against a total of 250 million in Yunnan and the rest of the riparian countries). The basin's inhabitants are, however, dismally impoverished, with an annual per capita income of no more than US$400.

There is evidently poverty amidst potential plenty. The inhabitants' livelihood hinges on the sustainable development of the basin's resources, 80 per cent of them being farmers and fishermen who are precariously dependent on the river for irrigation and fishing on the basis of fish stocks of over 1,000 species. Apart from power and food, the Mekong provides a relatively cheap means of communication, although, because of natural obstacles, it is not navigable throughout its length. The river is extremely versatile, and offers the potential of eco-tourism and flood control development. There is thus every reason to institute or restore governance in order to conserve and indeed enhance the productivity of the river basin's resources.

Public participation in the management of a river's catchment area may be looked upon as a desirable end in itself or merely as a means to an end. To some, it may be appealing as an end per se, not least because of its democratic nature; to others, it is a means to governance given the possible accompaniment to public participation of governance. Fortunately, public participation has been a traditional feature of the Mekong River basin and, although there is no necessary connection between public participation and governance, the two have frequently been found to coexist there.

The traditional participatory principle and governance

The Mekong River basin's inhabitants are by no means complete strangers to the participatory principle, which has been consistently practised in the basin's subnational communities since the distant past. This has fortunately existed in tandem with a pragmatic philosophy of governance, derived either from the Hindu and Buddhist principle of non-violence (*ahimsa*) or from the Taoist reverence for Nature (Wong 1999: 5–6). The stable concomitance of public participation with governance helped to maintain a sustainable ecology until the abrupt disruption caused by the advent of modern-day development, with its more exacting demands on resources. In fact, scrutiny of a handful of subnational communities in the Mekong River basin confirms the survival of the deep-rooted and robust participatory principle, evolved, as it has been, in the context of communal subsistence and cohesion. This holds true of the basin's wet and arid parts alike.

For instance, the controversial plan to construct the Kaeng Sua Ten dam in Thailand's northern Phrae province on the edge of the Mekong catchment area has for some time been the subject of a fierce national debate. Interest in the dam was revived in September 2003 on account of its claimed potential to avert persistent flooding of the country's central plain, which receives as much precipitation as does the Mekong River basin in the north-east (*Krungthep Durakit*, Bangkok, 19 September 2003, p. 11). A social impact assessment of the planned reservoir area reveals much that is invaluable in the indigenous system of natural resources management. People are said to use natural resources not only for subsistence but also for recreation and for spiritual and cultural activities. Their desire to protect the forest in the proposed dam area is in the interests of strengthening community ties, whose erosion, in the event of the possible loss of the local forest, would inflict immeasurable harm on them. Thus, left to their own devices, villagers claim to have set up their own rules prohibiting the community's members from felling even a single tree and have even helped to catch poachers and illegal loggers in the Mae Yom National Park in the projected dam area. The natural environment is seen to be indispensable for the community's cohesion, way of life and identity, thus post-impoundment resettlement of the dam area's inhabitants would, it is alleged in the study, entail disintegration of the community, because people would no longer be able to count on forest products for their survival; they would, on the contrary, experience severe hardship, their ability to adjust to a new way of life being impaired for good. Moreover, the villagers' age-old, tradition-hallowed knowledge about the forest and its rich biodiversity would, it is feared, be lost forever (*Bangkok Post*, 15 October 2000).

For centuries, the mountainous area of northern Thailand, which constitutes the headwaters of rivers traversing its agricultural central plain, has been dotted with small irrigation systems known as *muang faai* (diversion weirs) built and managed by farmers themselves (Sluiter 1992: 77). Traditionally, streams are dammed with a sturdy lattice-work of materials gathered from forests: rocks, hard wood, bamboo and earth. The dams serve to raise a stream's level sufficiently to allow its diversion into an irrigation channel that permits water to flow by gravity down to the fields; silt flows over and through the structure or is carried into the diversion channel to be deposited there. A similar system exists in Luang Prabang, the former capital of Laos, on the other bank of the lower Mekong (Sluiter 1992: 38). The *muang faai* has always been accompanied by a strict set of rules maintained by *muang faai* leaders, to ensure that the surrounding forest is safeguarded and the water distributed fairly to all members of the irrigation group. Recent changes brought about by imposed development projects have, however, threatened the viability of the traditional *muang faai* system. Since many of the forests have been, contrary to time-honoured tradition, logged over past decades, construction materials needed for the customary annual repairs are no longer readily available and free of charge. Moreover, as a result of large-scale logging activities, in the rainy season mountain streams become wilder and damage structures more frequently, while soil washed from bare slopes ends up clogging the channels. To eliminate the need for repairs, many farmers have unfortunately replaced the traditional structures with steel and concrete dams, which have the distinct disadvantage of not being as adjustable as the traditional dams; this is a problem especially where the forests have been cut down. Eroding soil and faster runoff can also cause erratic changes in streams and channels, demanding adjustments in dam height and channel maintenance.

Finally, the Mae Chaem district in Chiang Mai province in northern Thailand, which is situated just on the edge of the Mekong River basin and is surrounded by the formidable fortress of the Thanon Thongchai mountain range, had been blessed with well-preserved forests untouched by outsiders for many decades until the 1980s, when they were exposed to large-scale cash-crop plantation sponsored by international aid agencies in an effort to curtail opium cultivation on the highlands. Despite the best of intentions, the novelty led to extensive land-clearing, rapid deforestation, soil erosion and drought, and eventual abandonment of the programme that has left visible environmental scars. Faced with this, the official approach has been to put an end to the environmental drift by turning the remaining forest areas into national parks, wildlife sanctuaries and protected watershed areas no longer subject to human settlement.

As an alternative to the government's radical remedy, Care Thailand has launched an Integrated Natural Resources Conservation project aimed at broadening community planning by bridging the gap between villagers and government officers (Kungsawanich 2001: 1). Adopting a bottom–up approach by reinforcing time-hallowed community participation in natural resources management, whereby efforts are made to settle conflicts over the use of natural resources between ethnic groups and state agencies, Care has worked closely with *tambon* (sub-district) administrators in the project area. In retrospect, it was possible to pinpoint the mistakes of past top–down management imposed by international aid agencies. Contrary to previous experience, forest encroachment in the area occurred when villagers were dominated by profit-driven cash-crop plantation activities. Moreover, because mono-crop plantations consumed huge amounts of water, water wars between highlanders and lowlanders inevitably ensued. Instead of imposing a set of solutions on the communities, Care's renewed bottom–up approach has established village committees and mini-watershed networks to work out rules and activities for forest conservation. Although Care's approach is said to be bearing fruit in the form of the slow recovery of forest areas, the threat of future deforestation remains and there is a constant challenge to find a proper balance between economic gain and ecological well-being.

Following in Care's footsteps, in the neighbouring Mae Chan district in Chiangrai province, located at the northern tip of the Thai part of the Mekong River basin, a pilot project is being conducted under the auspices of Global Water Watch and jointly sponsored by two US institutions, Auburn University and the Heifer International non-profit agency, to demonstrate how to conserve water resources with the help of traditional participation. Since 2001, the selected village of 813 Akha hill-tribe households has planted trees on 48 hectares of land and is determined to increase this by 16 hectares every year. Moreover, the villagers have been trained to use basic equipment to test water quality in four streams, which are their main water sources for both drinking and farming. Fortunately, tests have so far found no chemical residues in the streams, and the villagers have been urged to avoid the use of chemicals in growing vegetables, which occurs mostly along their banks. Of course, this would run counter to the widespread use of chemical fertilizers by vegetable growers in the province, and it remains to be seen whether the pilot project is too ambitious to catch on (*Bangkok Post*, 13 August 2003, p. 5).

Delving deep into the past, the Khmer empire held sway over a substantial area of the Mekong River basin. The "hydraulic city", launched at Angkor by Indravarman I (AD 877–889) and completed by his successors, is believed by some to represent the ultimate in sophisticated water resources management, although its exact nature has been subject to heated academic controversy. It is possible that, quite apart from its purely hydraulic nature, it symbolized the oceans that surrounded Mount Meru, the Hindu counterpart to the Greek Mount Olympus, with the large and impressive temple pyramids dedicated to the royal ancestors being taken to represent Mount Meru itself (Higham 2001: 60). The "hydraulic city" consisted of a series of *baray*, huge and generally deep basins enclosed in brick or turf dykes probably serving as water reservoirs in an irregularly irrigated zone where 90 per cent of the cultivated area was at the mercy of the vagaries of the weather. These colossal *baray* may have stored as much as 40–70 million m^3 of water (the same amount as the Cambodian Tonle Sap or Great Lake normally contains today) and irrigated a total area of 70,000 hectares. Although these structures may be said to symbolize an exceptional concentration of power (de Sacy 1999: 43), there must have been, at the grassroots level, public participation in their management by default, because the government machinery was obviously inadequate to deal with the nitty-gritty of the end-use to which such a massive amount of water might be put.

Arid north-east Thailand, where the bulk of the Mekong River basin in Thailand is located, is no less well endowed with structures to store irrigation water. It is dotted with counterparts to the *muang faai*, known as *thamnob* (dyked water tanks or ponds), which direct water flows from river tributaries to provide irrigation water for the rice crop. The old theory that the *baray*, found generally in other parts of the Mekong River basin, including in Cambodia, were the source of irrigation water is rejected by some on account of the absence of water outlets from them. In north-east Thailand, farmers are still using the ancient *thamnob* irrigation system, which may date from the Khmer empire and is currently maintained with full public participation, despite the fact that rice-growing centres moved to the deltas in the nineteenth century owing to the importance of the emerging maritime trade and the imperative need to minimize overland transport costs. As in the cases noted earlier, the *thamnob* irrigation system has made for sustainable development, in contrast to big dams, which have threatened the environment in the region (Ekachai 2003: 1).

The traditional participatory principle, unlike its modern-day counterpart, has been essentially non-aggressive, non-assertive, inward-looking and non-confrontational. Moreover, projects in which the principle has been applied, unlike their modern-day counterparts, are modest in their dimensions, not sophisticated in the technology adopted and of purely local interest; moreover, their nature and scale are not such as to generate any possible conflict between national and local interests causing the inhabitants to choose between the two competing loyalties. In particular, traditional participation had its early beginnings in an environment where, in conformity with Hindu and Confucian concepts, there was deference to authority and the ruler was believed to be benevolent. Such concepts have stood the test of time in the Mekong River basin. Whereas the Hindu idea may be territorially confined, the Confucian concept of *tian sia* may be said to correspond to contemporary global governance (Bell 2003: 58). If the traditional participation is still practised in subnational communities, it is as much a relic of the past as a defensive measure against the inroads of modern-day development. In fact, for reasons already noted, modern-day participation is a far cry from traditional participation. Nevertheless, transition to its modern-day version is perfectly possible, provided that any unnecessary baggage from the past can be jettisoned and fresh elements of modern-day participation are absorbed to take their place.

Indeed, traditional public participation in water management has been more prevalent than may appear at first sight (Chomchai 2005). In view of the fact that most water sources managed by local inhabitants have been found to be in better working order than are government-run small water sources, it is envisaged that the government will soon transfer the power to manage small water sources to grassroots-level administrative organizations as part of the general programme of decentralization or devolution (Ruangdit and Theparat 2003: 3). It is tempting to argue that the coexistence of the participatory principle with green ideology constitutes a strong case for requiring public participation in water resources development. There is, however, no assurance that the indigenous green ideology is sufficiently robust invariably to make for sustainable ecology and to resist the temptation of the prospect of economic gain, albeit short-term in nature, particularly when there is a need for communities not only to balance ecological well-being against economic gains but also to avert conflicts over the distribution of such gains, which could easily get out of hand and destroy traditional communal cohesion.

That the indigenous Mekong participatory principle should favour governance is intriguing from a public finance analytical standpoint. For one thing, governance is a "public" good from which a potentially infinite number of people could benefit simultaneously (it being impossible to prevent anybody from so doing), but which, because of "market failure", cannot be left to the market mechanism to provide on its own. Of course, the free riders' quality of life cannot but benefit from governance, although these culprits persist in plundering such key elements of the environment as the forest or wildlife for their own private gain. For another thing, governance, like insurance, may be seen to be a "merit" good, to which people tend to attribute insufficient merit. It may, however, be said to represent a new breed of merit goods since, in contradistinction to such classic cases as housing and insurance, the "merit" want it is intended to meet is imposed not from above but from below. The aim is to preserve the livelihood of the common people, which is threatened by an absence of governance, especially in public sector projects. Because the government appears to lack the political will to tackle environmental deterioration and the workings of its machinery are thwarted by "government failure", the ordinary people may be said to be playing an avant-gardiste role in environmental governance.

Public policy, economic growth and environmental degradation

Property rights refer to a bundle of entitlements defining an owner's rights, privileges and usage limitations in relation to a resource. These rights can be vested either in individuals, as in a capitalist economy such as that of Thailand, or in the state, as in fully centrally planned economies or those in the process of transition such as those of China, Myanmar, Laos, Cambodia and Viet Nam, which occupy the rest of the Mekong River basin.

There is room for the existence and indeed coexistence of a variety of property rights systems in both types of economic organization, for private property is not the only possible way of defining entitlements to resource use. If one adopts the classification system presented by Bromley (1991), one can envisage other possibilities, including state property or *res communes* regimes (where the government owns and controls property, *res communes* being publicly owned things), common property regimes (where property is jointly owned and managed by a specified group of co-owners) and *res nullius* regimes (in which no one owns or exercises control over the resources, *res nullius* being ownerless things). All of these systems create rather different incentives for resource use (Tietenberg 2003: 70).

State property regimes exist not only in socialist countries but also, to varying degrees, in virtually all countries in the Mekong River basin and elsewhere in the world. Parks and forests, for instance, are frequently owned and managed by the government in capitalist as well as in socialist countries. Problems with both efficiency and sustainability could arise in such regimes to the extent that the incentives of the bureaucrats who implement and make the rules for resource use diverge from the collective interest. Entitlements to use common property resources may be formally protected by specific legal rules or informally protected by tradition or custom. Common property regimes exhibit varying degrees of efficiency and sustainability, the actual outcome in a particular case being dependent on the rules, which emerge from collective decision-making. Although there are some successful examples, such as the system of allocating grazing rights in Switzerland, unsuccessful ones are much more common (Tietenberg 2003).

In several countries of the Mekong River basin, the bulk of property is publicly or communally owned and property rights are rarely clearly defined or strictly enforced. The stock response to this is that market prices need to be corrected. However, this requires that these countries have at their disposal an appropriate regulatory and institutional framework to internalize negative environmental externalities. Their governments' inability to administer and enforce the laws that are intended to correct such externalities means that market failure tends to persist. These are precisely the countries that can least afford to protect the environment. Even when they attempt to protect the environment or conserve resources, regulations are inconsistently applied and regulatory agencies are too poorly staffed and poorly informed to be able to monitor happenings and implement regulations effectively. The ultimate effect has been rapid degradation of valuable environmental assets as a result of profligate and random land-clearing, irresponsible farming practices and excessive water and air pollution. The situation is likely to persist unless some means are found to eliminate the institutional weaknesses that are at the core of the problem, that is, to define and enforce clear rights of access to and use of resources for producers, consumers and government alike so that such resources may be prudently used. Of course, this does not necessarily mean, as has been noted above, that these countries need to resort to private ownership of resources, which is anathema to socialist systems. On the contrary, effective property rights systems could take several forms; what matters is that governments match property tenure laws with the relevant social context (Hussen 2000: 412–413).

Res nullius property, or open-access resources systems, can be exploited on a first-come, first-served basis, since no individual or group has the legal power to restrict access. They are thus characterized by non-exclusivity and divisibility. Non-exclusivity implies that they can be exploited by anyone, and divisibility means that the capture of part of the resource by one group reduces the amount available to the other groups (Tietenberg 2003: 71). Open-access resources systems thus tend to create two kinds of externalities: a contemporaneous externality and an intergenerational externality. The former, which is borne by the current generation, involves the over-commitment of other resources to tapping the resource at issue, whereas the latter, to be borne by future generations, occurs because overexploitation of the resource reduces its stock, which, in turn, lowers future profits from the activity (Tietenberg 2003: 289–291).

The demarcation line between communal ownership and open-access resources systems may, however, not be as clear-cut as it appears at first sight, since sometimes a blanket term such as "common property ownership" is used to describe a situation where there is no private ownership. This somewhat hazy viewpoint gives the impression of orderliness rather than chaos. It is true that, theoretically, open access could lead to a "tragedy of the commons": because everyone has access, all have the rights to the resource and its scarcity value is ignored. On the other hand, it is argued that the reality is that most commons have a property rights scheme, either formal or informal, that works to allocate resources in a more economically efficient manner. Indeed, there are said to be numerous documented examples of self-governing commons in which people work as a collective unit and respect the scarcity value of the resource at issue. These groups are believed to succeed because they establish common property rights that include sharing rules, exclusion principles and punishment schemes. Government intervention is one way of forcing members to cooperate; but this is not necessarily the only way. Members may also be able to cooperate by agreeing to abide by the decisions of an external regulator, who can be appointed by them (Hanley et al. 2001: 131, 156; Ostrom 1990).

It is not uncommon to hear that the source of contemporary environmental problems in a capitalist economy is the market system itself or, more specifically, the pursuit of profit. Those who espouse this view look longingly at centrally planned economies as a means of avoiding environmental excesses. However, centrally planned economies have not, historically, been able to avoid such excesses either (Tietenberg 2003: 62).

Of course, property rights regimes do not, by their nature, function *in vacuo*; they need a sturdy state infrastructure to sustain them. One thing appears to be clear, however, from the state's management of property rights regimes in the Mekong River basin: because of government failure and in the absence of Ostrom's "external regulator", all *de jure* property rights systems other than private ownership have degenerated into de facto open-access regimes. One of the most unfortunate, but recurring, realities in some of the countries of the Mekong River basin, notably Cambodia, Laos and Myanmar, is political instability, which is one of the main sources of government failure. Internal strife sometimes erupts into prolonged political conflict and even civil war. In this kind of political climate it would be extremely difficult, if not impossible, to implement effective population and resource conservation policies based on long-term visions. Instead, public policies are conducted on a piecemeal basis and generally as a reaction to crisis situations. This entails an apparent lack of responsible stewardship of resources that are critically important to the long-term survival of a nation (Turner et al. 1993; Homer-Dixon et al. 1993). The imposition of strict self-discipline in pockets of governance, especially in small local communities, has fortunately been able, like the Swiss communal grazing rights, to curb the worst excesses of the de facto open-access regime; elsewhere, as might be expected, widespread environmental destruction has resulted.

When, in the past, the person–land ratio was still favourable, the regime performed perfectly well. When, however, it is suspected that public water resources projects are disrupting the natural river flow and starting adversely to affect the natural fish supply, or when too many people are exploiting dwindling resources, people do not hesitate to protest. "Robbing the rural people of their means of sustenance or traditional rights" then becomes part and parcel of battle cries against officialdom. The de facto open-access system prevalent in the Mekong River basin may thus be said to be conducive to the treatment of natural resources as either "free" or "public" goods. "Capitalists" and their rural proxies are lured into behaving as though such resources are freely available for the taking. At the same time, the government is seen to be duty-bound to expend taxation proceeds to ensure the ready availability of such resources for "public" use, and the conspirators act as "free riders" and contribute little or nothing to their upkeep.

The de facto open-access system being the order of the day in Thailand, the indigenous green ideology has in general been too feeble to withstand the onslaught of globalization, population growth and economic growth. In fact, over the past 150 years or so, an unholy alliance between export-led growth and population increase, coupled with population movement, industrialization and urbanization, could be seen to have conspired to wreak havoc, especially on Thailand's apparently robust environment. For instance, a recent research project has found that 15-year-old mangroves in Thai forests have had to absorb up to the equivalent of 94 tons of carbon per hectare – almost 20 tons more than mangroves in Japan (*Bangkok Post*, 7 October 2000). Economic growth, which might aptly be termed Schumpeterian "creative destruction" since it has generated an illusion of being creative, started in the mid-nineteenth century when Thailand was, *manu militari*, compelled to open up. In the initial stages of economic laissez faire, such demands as were made on the environment in the interests of promoting exports of primary products from farms and mines were not too exacting: the country's economy took time to transform itself from a closed, subsistence system into an open exchange system. With the adoption of partial planning in 1961, when the first six-year development plan was launched, however, such demands, propelled as they have been from above, have gone beyond the carrying capacity of the environment, no matter how robust it may appear to have been. Of course, the problems Thailand faces with hazardous waste disposal, soil degradation, water quality deterioration, chemical and radioactive poisoning, coastal and marine degradation and loss of biodiversity are in no way unique to Thailand and are shared by all riparian countries of the Mekong River basin. Faced with the grave consequences of the persistence of the de facto open-access system, some governments in the Mekong River basin, with or without the support of civil society, have been compelled to take draconian control measures as well as remedial measures, particularly in fisheries and forest resources.

The deplorable state of the environment in the Mekong River basin is reflected in the fate of the Mekong giant catfish, *Pangasianodon gigas*, the world's largest scaleless freshwater fish, which is capable of growing to more than 3 metres in length. Both commercial fishing and river development, especially the Mekong navigation improvement project involving the blasting of rapids, are threatening the life cycle and long-term viability of the giant catfish. The rapids and whirlpool system in the Chiang Khong–Chiang Saen reach (in Chiangrai province in northern Thailand) is believed to be the only area used by the species as its spawning ground, and normally fishermen in the locality, like their ancestors before them, lie in wait along the Mekong during the dry, spawning season in the hope of intercepting the fish in their 300 km upstream journey. Unfortunately, no giant catfish have been captured in Thailand since 2001 and they appear to be in danger of disappearing from the country completely. Faced with this dire prospect, the Mekong Fish Conservation Project was intended to protect populations of migratory fish in the basin. The project buys live fish from fishermen in Cambodia and releases them back into the wild after due weighing, measuring, extracting DNA samples (for genetic studies) and tagging. Fishermen are given incentive prices for their catches and also a small fee for returning tags from recaptured fish. The project was launched in Chiang Khong in Thailand in 2000 but, with the collapse of Thailand's giant catfish fishery in 2001, it has had to be moved downstream to Cambodia, which is the only remaining place where the fish are captured on a regular basis, albeit in declining numbers. In general, catches of the giant catfish have declined by 90 per cent from their 1983 level. The migratory study has conservation relevance since it demonstrates the importance of free-flowing rivers and the link between floodplain habitat such as Tonle Sap Lake in Cambodia and the spawning habitat of the Mekong River. The research team of the project hopes to help establish no-fishing zones in the Tonle Sap Lake, the Tonle Sap River and the Mekong River – a critical habitat for migratory fish. There is no doubt that the stakes are high: fishing is a way of life for the people of the Mekong River basin and the loss of fish species means that millions of people will lose their food security, livelihood and economic viability (Roach 2003: 3).

The Mekong River basin's loss of forest resources is as serious as the depletion of its fish stock. Thailand has less than 30 per cent of its forest cover left, and since 1989 has had a ban on logging in natural forests and has implemented a series of supporting measures to protect the remaining forest cover and to promote private sector involvement in forest management and plantations (FAO 2000). The mismanagement of Viet Nam's forests over a period of two decades has similarly forced its government to ban production from its forests. To reinforce national efforts, bilateral agreements have been concluded, especially with regard to joint border areas. Thus Thailand and Myanmar have signed an accord to preserve 60,000 km^2 of fertile forestland in the Tenasserim range along their joint border, which is rich in biological diversity and is home to the world's second-largest wild tiger population after Siberia (*Bangkok Post*, 24 July 2003, p. 5). In most other countries of the Mekong River basin, natural resources extraction for export is still a key element of the development strategy. This suggests that commercialization or international trade is an important factor contributing to a rapid rate of deforestation (Rudel 1989).

Moreover, governments are frequently confronted with an urgent need to finance both domestic and international debt, which puts pressure on them to offer their natural resources for sale at a discount (Korten

1991). Thus in Laos, logging is a major export earner: wood products account for more than 35 per cent of its export proceeds and the share of forestry in the country's gross domestic product (GDP) is estimated at about 15 per cent (FAO 2000) – and much more when illegally exported timber is taken into account (Ojendal and Torell 1997). In a similar vein, the forestry sector in Cambodia makes up about 12 per cent of the national product and illegal logging and cross-border smuggling take place regardless of traditional tenure systems that make for environmental governance, as in the case of subnational communities in Thailand.

Of course, the export prices of such forest products are far below their full costs, which can be calculated only with "ecological pricing". For example, although reliable data are not available, the price of Myanmar's teak does not reflect, *inter alia*, the costs of the flooding that rapacious teak logging has caused to the country. Under-priced resources tend to be overused and depleted too fast, to the detriment of longer-term welfare (Berg 2001: 535), and unfortunately some governments in the Mekong River basin appear to have taken action too late to nip the deforestation problem in the bud. Despite austere control and remedial measures, the overall picture of deforestation in the Mekong basin is bleak, not least because of interdependence: illegal logging in Thailand, for instance, is checked to the detriment of the viability of the forest cover in Myanmar, Laos and Cambodia (*Bangkok Post*, 8 August 2003, p. 3).

Apart from property rights systems and measures of direct control and conservation, another relevant factor in the complex environmental situation of the Mekong River basin is fiscal policy. Although internationally accepted "polluter pays" and "user pays" principles may mandate some form of tax in cases where there are negative externalities, taxes are too frequently levied elsewhere in the economy in order artificially to lower people's costs of using scarce resources. The point is not that environmental issues call for more taxation but that they require specific forms of taxation that directly affect the individuals and firms that cause the environmental costs that society ultimately bears. Indeed, the imposition of environmental taxes and the elimination of subsidies that encourage environmental destruction or waste of scarce natural resources need not impose any additional burden on the economy. On the contrary, such taxes can alleviate costly tax burdens elsewhere in the economy and yield the multiple benefits of improved resource allocation, more resource-saving technological progress and reduced costs of other welfare-enhancing activities (Berg 2001: 538).

Such greening of taxation could be undertaken in two complementary ways. One consists in restructuring existing taxes in an environmentally friendly manner. Such an approach is aimed at modifying relative prices by taxing those products and activities that pollute relatively more than others. The other way is to introduce "eco-taxes". For instance, product taxes are applied to products that create pollution as they are manufactured, consumed or disposed of as pesticides and fertilizers, it being understood that such eco-taxes are implemented to tackle specific environmental issues on an ad hoc basis (O'Riordan 1997: 230).

Of course, a more comprehensive approach entails the greening of taxes in the broader context of tax reform. The proliferation of new environmental taxes and the restructuring of existing taxes raise the issue of the compatibility and coordination of these taxes with existing fiscal structures and policy. On the one hand, the compatibility of current non-environmental taxes with environmental goals needs to be reviewed; on the other hand, new eco-taxes must be properly integrated into fiscal structures.

A general explanation of environmental degradation in the Mekong River basin could go beyond government failure in the spheres of property rights systems and fiscal policy. Some scientists argue that massive environmental destruction is inevitable when the human population is expanding exponentially. Others emphasize that far too many new substances have been introduced into the environment before their impacts on other species, let alone ourselves, have been ascertained. Economists tend to argue that people are generally too greedy and short-sighted, whereas Marxists concentrate their attention on a subset of the human race, the capitalists, and agree with moralists in arguing that this class of people is particularly avaricious and myopic. The various disciplines, in isolation or in combination, tend to vaunt the explanation of environmental crises that is consistent with their pattern of thinking, and there is no particular reason, at least at the intellectual level, to take issue with any of these explanations rooted in individual disciplines of thought. Each explanation provides useful insights (Norgaard 1994: 65).

There is no doubt that such general explanations are quite relevant to the Mekong River basin. For instance, population explosion has been a clear concomitant of Thailand's economic growth, particularly in the period prior to the onset of the Asian economic crisis between 1988 and 1997, when double-digit growth placed the country in the league of the world's fastest-growing economies. Particularly in the past three decades, a substantial proportion of South-East Asia's impressive economic growth can be attributed to a "one-off fire-sale of natural resources", which means that it may be harder to grow as fast when the trees, the fish and the soil are depleted. A series of articles in the Bangkok *Daily Manager* (e.g. on 16 October 2000) lament the fact that planned development in the country has resulted in pauperization of the masses, whose fate has been sacrificed at the altar of "development". It is claimed, in particular, that, since the inauguration of the trend-setting first development plan of 1960, priority has invariably been given to the promotion of and assistance to the industrial and business sectors, in which politicians have obvious vested interests, at the expense of agriculture. It has been claimed that industrialization, be it in rural or urban areas, reflects a conspiracy on the part of politicians and business interests to ruin the farming class and turn it into an impoverished, landless proletariat.

Again, at a *Bangkok Post* seminar on the People's Agenda held in October 2000, a number of academics and activists committed to acting as spokespeople for the ordinary people have pronounced upon the government's performance record. From this viewpoint, most of the damage to the environment and rural communities has been inflicted by the government itself. In public sector projects in a typical top–down "development" programme, in which no public inputs were used, the government, itself a product of electoral politics mired with money and power, is said to have acted as an independent interest group unaccountable to people at the grassroots level. Such projects are said to have typically allocated resources, without consulting the localities concerned, to one group of people to the detriment of another; the rural sector is thus robbed of the resources necessary to sustain its livelihood.

Experience with "promoted" private sector projects in monoculture is said to be equally dismal: they are said to run counter to traditional norms in that they are chemical intensive in nature and have destroyed the soil, polluted the water and landed farmers in a mountain of debt. Degradation of natural resources and the resulting poverty stemming from both public sector and "promoted" private sector projects are predicted eventually to lead to the collapse of the countryside. In the meantime, lost in dire poverty and hardship, some villagers have been observed to resort to endless, frustrating protests, while others, particularly young ones, migrate to the cities in search of a better life. For its part, the ruling élite is said to have suffered from "intellectual bankruptcy", indiscriminately jumping on the corporate-led globalization bandwagon without realizing how this could harm the economy and the communities of the country. In sum, the government policy of industrialization – whether import substituting or export promoting – and export-led agriculture is believed to have resulted in a rapid growth of the urban sector while leaving the farming, rural sector bankrupt.

There is no doubt about the seriousness of purpose of the governments in their campaign to alleviate poverty, although cynics could well see it as an obvious vote-winning gimmick. The primary focus on increasing per capita GDP, particularly through increased capital formation, unfortunately has two major flaws as far as the environment is concerned (Hussen 2000: 411–412). First, the conventional measure of GDP does not take into account the depreciation of natural or environmental capital. Thus, a focus on blindly increasing GDP is likely to have a detrimental effect on the natural environment in the long run. Secondly, capital formation is traditionally conceived of in terms of large-scale capital-intensive projects such as dams, highways and factories, and these projects are generally implemented without an adequate assessment of their impact on the natural ecosystem (Goodland and Daly 1992). The upshot of such a pattern of growth is continued environmental degradation.

In the Mekong River basin, the economy is primarily agrarian and the environment is an important input into many production activities. Environmental degradation thus has an adverse effect on productivity, which in turn leads to a reduction in income. The important implication of this result is that poverty-alleviation programmes are likely to fail in the long run if their primary focus is on increasing GDP. Such a growth ideology undermines the economic significance of the natural environment, and protection of the environment should be an essential element of poverty alleviation (Hussen 2000: 412; Bandyopadhyay and Shiva 1989).

On the other hand, examples drawn from developing economies generally make it clear that the relationship between economic growth and environmental abuse may not be as straightforward as is frequently suggested by critics of economic growth. For instance, it is argued that one cannot simplistically assume that higher levels of production will be accompanied by higher resource use and therefore more pollution. There is no such clear-cut relationship. Higher per capita incomes and the higher levels of technology that accompany economic growth could also work to reduce pollution and the inefficient use of resources. In particular, it is pointed out that the overall level of economic institutions in the more developed economies permits these countries to create more effective incentives to reduce pollution and the waste of resources (Berg 2001: 542).

It is fair to conclude that economic growth can be both the cause of and the cure for environmental problems: growth increases our demands on the environment but growth also gives us the time and money to do something about its undesirable side-effects. In other words, even with economic growth, market failures still remain. Blindly promoting economic growth as the remedy for all ills is unwise; what is needed is growth with accountability (Hanley et al. 2001: 133).

It is comforting to note that economies can grow their way out of environmental problems. Thus the famous Environmental Kuznets Curve (EKC) states that, as per capita incomes grow, environmental impacts rise, hit a maximum and then decline. Economic growth results in an increasing use of resources and land clearance. In particular, if a country starts from an early development stage as an agricultural economy, then industrialization also leads to an increase in emissions, as manufacturing takes over from agriculture as the dominant economic activity. As incomes rise, there is a growing demand for environmental quality. This leads to an increase in government protection of the environment and increasing green consumerism. Technological improvements over time make production per unit of output cleaner. At the same time, changes in the structure of the economy, such as moves from manufacturing to services or high-tech industries, occur. In time, increasing scarcity of "environmental quality" drives up its relative price, and this means that less is consumed and more is preserved. Looked at in a dynamic perspective, the EKC hypothesis implies a hierarchy of needs to be satisfied: first people want food and shelter; later they also demand amenity. To put it another way, first of all we protect ourselves from Nature, and only later on do we protect Nature from ourselves (Hanley et al. 2001: 130–131).

Top–down public sector water resources projects

Traditional public sector river basin development has been no exception to the prevailing top–down trend of development; for it is commonly understood to mean the public sector sponsored construction of dams, reservoirs, weirs and irrigation infrastructure and the expansion of protected areas into upper water catchment areas to maximize the resource value of the system. This has evolved from four decades of experience, particularly in Thailand, during which river basin plans have not gone beyond an obsession with the public sector's creating large-, medium- and small-sized water storage, whether for flood control or for dry-season water use. Decision-making in such river basin development has typically been confined to a small group of technocrats, economists and irrigation engineers, together with foreign experts brought in by international and regional aid agencies such as the World Bank and the Asian Development Bank.

This situation may be said to have originated from three key factors in Thailand: centralization of the social and economic planning framework; an overdependence on dominant "expert knowledge" in river basin management; and export-oriented economic development efforts that tie production to the global economy. Under such centralization, irrigation legislation has consistently allowed the dominant Royal Irrigation Department (RID) to develop water resources unilaterally. Likewise, the relatively new National Economic and Social Development Board, the overall planning body, has undertaken to draw up development projects without any reference to or involvement of people living in the areas affected by the planned projects.

Traditional paternalism in river basin development dies hard; even in Thailand's current era of openness ushered in by the "people's constitution" of 1997, government approaches to river basin management continue to exclude popular participation or at best allow only "stage-managed" participation. Clear illustrations of the practice are provided by the management of the (Me)Kong-Chi-Mun diversion scheme in Thailand's north-east and plans to divert Mekong headwaters from the northern Kok and Ing tributaries into the Chaophraya (the country's main river feeding the central plain), via the Nan River (a tributary of the Chaophraya), two of the schemes that are likely to affect the Mekong mainstream. Moreover, direction by "experts" goes hand in hand with centralized river basin planning. Apart from the imposition of technological solutions, whether in the form of large dams or smaller structures, people are also confronted with the loss of the status and value of their own water management knowledge.

Despite the multitude of problems that public sector water development projects have created for effective water management, there are only limited opportunities for effective public challenges to such "mainstream" thinking and dominant knowledge system. As late as March 2003, the Thai government was still displaying its reluctance to give up its monopoly on water use (Wangvipula 2003: 6). Critics (notably Chantawong 2002: 2) point to the fact that, although dam construction in north-east Thailand, which constitutes the bulk of the country's Mekong River basin, had already given rise to problems of soil salinity, plans for further projects continue unabated. Again, despite the fact that dams have been unable to solve flooding or water storage problems, society continues to place its faith in the experts' rationale for further dam construction, and the logic whereby water cannot be allowed to flow wasted and unused into the sea apparently continues to underpin further dam construction and water diversion projects.

Finally, economic reorientation appears to have changed the system of values associated with managing water. From a pure public good, managed and used communally, water has become a kind of input into mass-production agriculture geared to the requirements of the world rather than the domestic market.

That the top–down, paternalist approach is typical of river basin development is borne out by evidence from other parts of the Mekong River basin. Dr Yu Xiaogang, director of Green Watershed, a non-governmental organization (NGO) in Yunnan, one of China's principal riparian provinces of the Mekong River, told a seminar held in Ubon Ratchathani (in north-east Thailand) in November 2002 that the Chinese government had not given the public a role in decision-making with regard to the cascade of dams under construction across the Chinese segment of the river. According to Dr Xiaogang, the local community had not only been denied access to economic impact assessments (EIAs) but also had to make sacrifices, purportedly for the sake of the "national interest", through being subject to forced resettlement away from their homes and having to pay more for power – despite their relative proximity to the Man Wan hydropower dam – than people in Guangdong on the east coast, 1,400 km from Yunnan. Dr Xiaogang was aware that it is not only the local interests of Yunnan but also those of downstream countries that are affected by Chinese river works upstream. In retrospect, it was argued that the dam had been meant not so much for the local villagers as for urban and industrial sectors located elsewhere (Panwudhiyanont 2002: 28; Sakboon 2002: 8A).

Although obsessed with the question of how best to develop the river's resources, the Lower Mekong riparians have equally been kept in the dark about upstream developments. It could be argued that, had they been more resourceful and more successful in overcoming their mental block, they could have been more active in trying to glean crucial information from Chinese sources. It is true that Chinese sources have been difficult to come by, and the more accessible Western sources had to wait until the early 2000s to blow the whistle on works on the Lancang. The lower riparians were living in a fool's paradise until, in the early 1990s, they were jolted by the impact of the upstream regulation. In the new circumstances of a regulated river, the Mekong River basin's inhabitants south of the Chinese border were brought face to face, for the first time, with the hard fact of their vulnerability, since, in this part of the Mekong, the Upper Mekong's contribution to the river flow is not 18 per cent, as in its distributaries that constitute the Mekong River's estuary, but virtually 100 per cent.

The Upper Mekong cascade of hydropower dams is now becoming a reality, the 1,500 MW Man Wan dam (994 metres above sea level) having come on stream in 1993. Its storage capacity of about 10 million m^3 is comparatively limited (compared with the 247 million m^3 planned for the 5,000 MW Nuozhado dam at 807 metres above sea level, which will possibly be the largest dam of the cascade), and it was followed by the 1,350 MW Dachaoshan dam, at 895 metres above sea level and with a slightly smaller storage capacity (9 million m^3).

Despite Man Wan's modest dimensions, it did not take long for its appreciable negative impact to be felt on downstream river flows. By the dry season of 1995, the river was at a record low level and had begun to give the lower riparians an inkling of the shape of things to come. It wreaked havoc with the routine activities of Thai and Laotian inhabitants alike. Thai touring boats were unexpectedly stranded at the Golden Triangle. Thai farmers in the vicinity found it more difficult to lift the river water to their fields, and Laotians, on the other bank of the Lower Mekong, were prevented from holding their traditional annual aquatic festival at Luang Prabang, their former capital. It eventually transpired that works at Man Wan had required the temporary diversion of the Upper Mekong. A complaint is reported to have been lodged by the governor of the affected Thai province (Chiangrai) with his Yunnan counterpart, who, although acknowledging that some tunnelling had necessitated the temporary curbing of the river flow, refused to agree to notify his Thai colleague if such works were planned in the future (Hinton 2000: 18). It was clear that the Chinese were unwilling to impose upon themselves even the moral obligation, not to mention obligation in international law (Toope 2001: 104), of notifying the lower riparians of planned upstream activities.

More generally, according to the November 2002 seminar in Thailand (Panwudhiyanont 2002: 28–30), the Man Wan dam has had a serious adverse impact on areas immediately downstream of China. After its construction, the river's hydrological pattern is said to have undergone a radical transformation: the water level paradoxically rises in the dry season but falls in the wet season, and there is a disturbing uncertainty about it at any given time, since upstream release for hydropower generation is the determining factor. In fact, the water level of the Mekong River below China is now at the mercy of the upstream Chinese, and it has been subject to unnatural fluctuations (*Daily Manager*, Bangkok, 10 January 2003, p. 13). In particular, northern Thai fishermen were faced with an unusually low level of the river in 1994 when Man Wan picked up steam. Uncertainty about the expected water level is said to have left fisheries in disarray since fishermen can no longer "read" the water level and pick out the right fishing equipment from the panoply at their disposal. Because the water level could be metres off the mark, fish, especially the giant catfish, an obviously endangered Mekong species, could easily take evasive action. The Mekong River has also become increasingly turbid, which affects fish stocks that are sensitive to changes in the water quality (*Daily Manager*, Bangkok, 10 January 2003, p. 13). There is said to have been a negative correlation between the Mekong's hydrological upheaval and the size of the catch. In fact, since 1994 the number of giant catfish caught in Thailand each season has dwindled to 0–8, as against an average of 40 during the preceding eight-year period. Fishermen believe that reefs and shoals just above the Golden Triangle are the species' central spawning ground and rock-blasting spearheaded by China to make way for safer navigation has sounded the death knell for the unique species. Chinese vessels of up to 500 tons are now able to ply the Mekong, which is a more attractive route than the more costly and time-consuming overland route to the South China Sea (Kongkrut 2003: 10). This seems also to be well in keeping with China's relentless push towards the sea-lanes, not least to be assured of its oil supplies (Hayes 2001: 21). The reduced opportunities for fishing are also said to entail the irreparable loss of a culturo-political heritage: it has suddenly dawned on Thai and Laotian fishermen on both banks of the Mekong, who have, for centuries, stood shoulder to shoulder in managing shared natural resources, that they are no longer in command of their age-old heritage.

Indeed, the downstream impact of upstream river works is far more extensive than appears at first sight. When the Man Wan reservoir was filling up in 1995, even the water level in Cambodia was appreciably affected. In particular, the floods of 2000 are said to have been the worst for two decades, wreaking havoc on urban centres from Chiang Saen (Thailand) to Luang Prabang and Vientiane (Laos) to cities further downstream in Cambodia and Viet Nam (Kongkrut 2003: 10). Villagers were accustomed to living in harmony with nature, including natural flooding. Upstream river works are believed to have changed all this, and the time-hallowed harmonious relationship between humans and Nature has unfortunately been turned into an adversarial one.

Perhaps more disastrous in its downstream impact than China's Man Wan has been Viet Nam's Yali Falls, built on the Se San River, a major tributary of the Mekong River, about 80 km above its border with Cambodia, where the Se San flows into the Sre Kong River before the latter's confluence with the Mekong. With an installed capacity of 720 MW – less than half the size of Man Wan – it is nevertheless the Lower Mekong basin's largest dam, on one of its largest tributaries. It is alleged that construction of the dam was started even before agreement to the EIA had been forthcoming (Sakboon 2002: 8A). Ironically, the EIA, conducted by a Swiss company, maintained that people downstream do not depend on the river and that the dam would therefore have no impact. In any case, civil groups at the November 2002 seminar in Thailand put forward the view that an EIA is just a tool amenable to government manipulation, and, as such, too frequently fails to consider and evaluate the needs and rights of the affected communities (Sakboon 2002).

No sooner had power generation at Yali Falls begun in 1998 – after five years of construction – than the irregular releases of water from its reservoir led to a radical alteration in the hydrological regime and the water quality of the Se San River downstream. Unusual and dramatic fluctuations in river levels along the Se San are said to have had major environmental and socio-economic impacts downstream in Ratanakiri province in north-east Cambodia (affecting more than 20,000 inhabitants). Severe flooding in the early days of hydropower generation is said to have drowned 32 Cambodians, and flash flooding has turned out to be a hazard even in the dry season. Forced evacuation is said to have taken place after the Yali Falls impoundment, and much has been lost from unexpected flooding (Panwudhiyanont 2002: 30). On the other hand, river levels were particularly low in the dry season of 1997 and 1998. Moreover, there has been no dearth of reports of serious illness among humans and livestock alike, with river water apparently contaminated with the detritus of the corroded riverbed and banks as a result of upstream impoundment. Food security and nutrition are also said to be jeopardized, as people have suffered the irreparable loss of their lowland and swidden rice crops through irregular flooding, in tandem with a waning fish catch on account of a sharp decline in the fish stock, their main source of animal protein. Apart from giving confusing signals to migratory species, the Se San is said to have been blighted by increased turbidity, which has adversely affected all fish species to the extent that some have disappeared from the river altogether. In sum, the dam has precluded people from harvesting their traditional resources from the river, and they have been forced to fall back on the already overexploited terrestrial resources in order to survive. What is no less disturbing is that, despite the devastation that Yali Falls has wrought on downstream areas, Viet Nam plans to build two more dams (Sesan 3 and 3A) downstream of Yali Falls (Sakboon 2003: 8A). The familiar tale of woe is indicative of the shape of things to come, with the prospective completion of the entire Chinese cascade on the Lancang.

From the standpoint of realpolitik, the desire to demonstrate good neighbourliness is too feeble a force to bring about a closer rapport between the Upper and Lower Mekong basins or between countries in the Lower Mekong River basin, even when China and Viet Nam were at pains to secure the goodwill of their neighbours for possible deployment in the international arena and even when the signing of the Mekong River Commission (MRC) agreement in 1995 entailed the need to meet certain obligations vis-à-vis one's neighbours. A sagacious balancing of upstream and downstream interests should, however, propel things in the right direction, although there may not be too many areas where such interests happen to coincide – eloquent examples are China's interest in navigation of the river and its southern neighbours' corresponding need for data on the Lancang for studies of and operations on the Lower Mekong. Thus the sharing of data on the river has taken place, for which China signed an agreement with the MRC in April 2002. Under the agreement, China and the Lower Mekong basin countries will exchange hydrological data, which should enable the lower riparians, *inter alia*, to forecast flooding better in the wet season and, if dry season data were also forthcoming, to forecast low flows in the Lower Mekong. The acid test of whether the exchange is an "equal" one has not been slow in coming: barely four months after the signing of the accord, parts of the Lower Mekong basin were abruptly submerged under record high levels of the river. Only an *ex post* evaluation could indicate whether data from China were provided sufficiently promptly to allow the lower riparians, especially the Cambodians and Vietnamese at its tail end, to take preventive and relief measures. For its part, China needs water-level data for the purposes of navigation in the Lower Mekong, the only realistic alternative outlet for cheap industrial goods from Yunnan being the more costly overland route to its eastern coast. Of course, the Salween River through Myanmar, China's traditional ally, is another alternative, though it may not be as economically palatable as the Lower Mekong.

There is no doubt that the Chinese have pursued their navigation ideal with an unprecedented single-mindedness. A Chinese engineer from Yunnan is reported to have stated that China will achieve what has been an impossible dream for France: to navigate uninterrupted from China to the Mekong delta in Viet Nam (Tangwisutijit 2003: 5A), regardless of the costs imposed on downstream neighbours.

The emergence of modern-day public participation

Although public involvement in river basin planning would have been unlikely to have prevented the negative downstream impact of upstream river works, access to information on public sector dam construction could at least have given people warning of the impending disaster. The behaviour of upstream countries is in breach of such internationally accepted norms of governance as information disclosure and transparency. With traditional paternalism being so persistent, it is utterly unrealistic to expect public involvement when water resources projects of transboundary dimensions are at issue, and other factors such as a basin state's conviction in its absolute territorial sovereignty also come into play.

As a belated reaction to this paternalistic, élitist and externally oriented pattern of river basin development, over the past decade popular scrutiny of river basin development projects appears to have emerged, albeit domestically, particularly in Thailand. This scrutiny has concerned both the projects themselves and the process. On the substantive side, salient issues are believed to have included the efficiency of dams and irrigation structures, environmental and social assessment, economic efficiency, RID's water allocation principles, compensation mechanisms for those adversely affected by projects, and water demand forecasting. On the other hand, process concerns are believed to have included overly centralized state-centric decision-making systems, the inability of people to gain access to and involvement in decision-making at all levels, and the absence of opportunities for community-based knowledge to be employed in river basin development (Chantawong 2002: 2).

In general, modern-day participatory principle and practice may be said to have arisen out of the growing recognition of the inadequacies of the top–down approach, particularly in rural development, as exemplified by its failings in the case of Thailand (see above). By the 1980s it had become clear, at any rate in the academic literature, that externally imposed and expert-oriented forms of development could no longer be tolerated. Chambers (1983, 1994) was one of the most influential writers putting forward participatory approaches that would make the people's involvement central to development.

As elaborated elsewhere (Chomchai 2005), the modern-day participatory principle in Thailand may be said to have its origins in the 1932 transformation to a limited or constitutional monarchy. Apart from constitutional checks and balances intended to steer the country clear of what is known in South-East Asia as "money politics", stamp out corruption and break the hold of money barons on politics, civil rights and liberties are augmented so that they may come to life with popular participation. As spelled out in the Information Act of 1997, these rights include access to information that is in the public domain and is in a government entity's keeping. In particular, the constitution guarantees local participation in environmental protection in such a way that indigenous communities are accorded the right to take part in the maintenance and management of natural resources and the environment and to demand information, clarification and justification from a government entity before it proceeds to approve, license or carry out a project that has an impact on the environment, their health and hygiene. What is more, any activity or project that might seriously affect the quality of the environment is prohibited unless an environmental study is undertaken that is endorsed by independent agencies, including representatives from environmental NGOs and university academics. In other words, internationally accepted norms of governance have been enshrined in the new Thai constitution.

Other parts of the Mekong River basin similarly espouse at least the principle of public participation. Although one person cannot pretend to speak for the Mekong River basin in China, the principle appears to be accepted there. It is said that, because local-level management challenges cannot always wait for national institutions to deal with them, local governments and local people should be encouraged to manage their own environment. China's national interests do not necessarily reflect those of Yunnan and Yunnan's interests may not be of high priority to China's central government (Ting 2001: 33). In Laos, the constitution is the product of discussion by the people throughout the country. There are four guaranteed levels of public participation in public sector projects: information-gathering, information dissemination, consultation and participation (Environmental Research Institute 2001). In Cambodia, it is admitted that the government has indirectly delegated a certain amount of authority to civil society (Hourn 2001: 92). In Viet Nam, the requirements for public participation in environmental decision-making are four-fold: knowledge, participation, discussion and control (Can et al. 2001: 21).

It is thus clear that, whether or not public participation is constitutionally guaranteed, most countries in the Mekong River basin recognize, at any rate in principle, the significance of public participation and its key elements. However, such recognition is, by its nature, territorially confined. The next step is, of course, for these basin states to agree to the transboundary extension of public participation. Exploration of the involvement of the public and public opinion in the work of the Mekong River Commission, whose raison d'être makes its activities transboundary in nature, was initiated in 1996, barely a year after the MRC's foundation, and public inputs are believed to be required, to begin with, at various stages of the formulation of the Basin Development Plan. This could, however, cut both ways. On the one hand, the public and the MRC could learn how to reach an accommodation. For its part, the MRC would be brought to the realization that it can no longer operate *in vacuo*, and the public could put their familiarity with the traditional participation principle to the test. The Basin Development Plan is an excellent candidate for launching the proposed innovation, because such general principles as sustainable development, and the governance underpinning it, would not arouse controversy. In any case, general principles and planning for the basin are too far removed from day-to-day livelihood concerns for the basin inhabitants to fret over them. On the other hand, the crunch will come when project-specific issues, especially in their locality, are debated either in isolation or with reference to the national interest of a particular basin state or indeed of the entire Mekong River basin. No amount of goodwill accumulated from accommodation over the Basin Development Plan would be able to avert a possible showdown when conflicts of interests and loyalties threaten to cloud the judgement of basin inhabitants.

A collective stance, not only on the Basin Development Plan but more generally, is proposed in the MRC Secretariat paper on Public Participation in the MRC (1999), in which all four stages mentioned in the Lao constitution are enumerated, albeit with a difference: the stages are intended to culminate in some decision-making power. If this is accepted by the MRC member countries collectively, it would be a heroic transboundary step, although it will not be able to bridge the outstanding divide between the Upper and Lower Mekong basins.

On a more universal plane, the need for public participation has more recently been recognized in some international instruments (Benvenisti 2001: 118). An obvious prerequisite of public participation is the dissemination of information by government. A government's duty to disseminate information also finds support in international instruments related to international common-pool resources (ICPRs) (Benvenisti 2001: 117). As has been noted in connection with the Lao constitution and the MRC Secretariat paper, modern-day participation requires much more than mere dissemination of information. It is not expected simply to be an exercise for show, where people are allowed to have a say, only to be forgotten later. By its very nature, participation cannot be a one-off exercise, but occurs throughout the life cycle of a project or programme. Indeed, the right to participate involves not just freedom of speech but must delve comprehensively into specific issues that will directly or indirectly affect the lives of those involved. These include such things as the right to determine/negotiate compensation if it is to be made, to determine/negotiate a changing way of life that may take place, the right to determine/negotiate property rights, and the right to be informed about the degree of risk that people may incur (Turton 2000: 28).

The whys and wherefores of public participation are frequently expressed in terms of efficiency and righteousness. Public participation could be made more effective and less costly, particularly in small-scale institutions, which are likely to be more sensitive to the concerns of those directly affected by the uses of such ICPRs (Benvenisti 2001: 124) as are represented by the resources of the Mekong River basin. The existence of a number of relatively small institutions, each responsible for a single sub-basin, could facilitate efficient intra- and inter-basin trade in the resources, with a central institution in the form, for instance, of a national Mekong committee serving as a forum for negotiations and even as a clearing-house for transactions among sub-basin representatives. In an effort to bolster efficiency with considerations of human rights and group rights, one can trace in international law an increasing recognition of the claims by minority groups, especially indigenous peoples, to a right to manage the natural resources in their vicinity autonomously as part of their claim to self-determination and cultural protection (Benvenisti 2001: 125).

Whatever terminology is used, theories of public participation all subscribe to the merits of participation by stakeholders who actually inhabit the river basins in question and those whose livelihoods depend on their resources (Editorial 2002: 1; Chantawong 2002: 3). Owing to the prevailing enhanced degree of integration and interdependence of modern economies, the latter sub-group of stakeholders can be very large indeed. As the MRC Secretariat paper of 1999 is at pains to make clear, because of the transboundary nature of the impacts of river works, stakeholders could well include people living in countries outside the Mekong River basin.

There is justifiably a feeling of apprehension that the embedding of civil society, seen as the foundation stone of public participation, in the Mekong River basin may not be as simple as it appeared at first sight. Civil society is rooted in specific social, economic and political contexts, and civil society in river basin development is no different from this. To transfer the participatory principle from one context to another is as difficult as implementing a physical design structure out of its intended context (Editorial 2002: 1). It is true that there is in the Mekong River basin a solid groundwork of traditional participation and that the basin's inhabitants have an intimate familiarity with their part of the catchment area. Nevertheless, academics and social activists have recently jumped on the bandwagon of modern-day public participation in tandem with civil society and have attempted to force its practical implementation on the basin's inhabitants. Without proper grassroots orientation and appreciation, this smacks of the imposition of apparently alien institutions on unwilling basin inhabitants, and this attempt could well backfire.

Be that as it may, the real substance and significance of public participation are believed ultimately to hinge precariously on de facto power relations (which could depart from *de jure* ones) among the various institutions and groups in society (Chantawong 2002: 3). There is thus no guarantee that the formal openness of the new constitution in Thailand or guarantees made in black and white in other parts of the Mekong River basin will filter through, because general pronouncements of principle are far removed from the operational level of river basin planning and management. A real participatory process for popular or civil society involvement involves challenging the existing de facto structures of river basin planning authority, as manifested in the over-centralization of power, reliance on "mainstream" knowledge systems and vested interests in river basin planning (Chantawong 2002).

Whether the right to such participation will be handed down as a matter of course by "mainstream" institutions remains to be seen. Hopes are high that the MRC's decision-making process will, as was made clear in the 1999 Secretariat paper, take into account the interests of civil society. Since 2002, civil society representatives have been invited to attend the MRC's Joint Committee and Council meetings, albeit as observers. The outcome of this gesture has been to let people down, according to a declaration at the November 2002 seminar in Thailand by representatives of "the local communities of the river basins in Thailand", which maintained that the MRC agreement "excluded local communities from making decisions about the Mekong River Basin and development" (TERRA 2002). However, actual power does not rest with the MRC nor has the MRC been invested with any supranational authority. If participation is unlikely to be handed down owing to the rigidity of existing power relations, it may perforce have to come from genuine, serious and persistent efforts by civil society (Chantawong 2002: 3), and ultimately even open confrontation and social unrest and instability.

Even where public participation is guaranteed by a Mekong River basin state's constitution, there has generally been great reluctance on the part of the authorities to allow it to operate on a consistent basis. Even in democracies, the tradition of paternal government is too well entrenched. Where public participation is allowed, then it is treated on an ad hoc basis and not as a general rule. Thus the Thai government's plan to allow people who live in degraded mangrove forests (which are closed to outsiders) to stay put under strict regulations, while helping with the salvaging of coastal systems, sees the role of the grassroots as a watchdog being valuable in contributing to ecological recovery.

Despite constitutional guarantees, totalitarian regimes generally do not willingly allow public participation to take place in practice, and such regimes operating in the Mekong River basin are no exception. People are, as a rule, kept in the dark about the government's intentions. As has been noted, China did not ask the inhabitants of the Man Wan area to participate in decision-making on the dam and they were confronted with its phased completion in 1993 and mid-1994 as a fait accompli. In a similar vein, the second dam on the Lancang, Dachaoshan, was built in 1997 and came on stream in 2001 without people in the locality having been consulted. Of course, the official newspaper, *People's Daily* (Wongruang 2002: 8A), keeps people abreast of such developments, but many of the basin's inhabitants are illiterate or have no ready access to the government newspaper. Nor is a newspaper the best medium for disseminating information about water resources development projects among stakeholders.

Transboundary cases are not handled any better than purely domestic ones by such regimes. Thus, despite the fact that the Cambodian government had been informed by Viet Nam that Yali Falls would be constructed, it failed to warn its own people downstream of the impending ecological disaster. Exceptionally, however, the affected inhabitants are consulted, although it is suspected that the public hearings are stage-managed and serve merely as a window-dressing exercise for donors' consumption. Thus, the Nam Theun II hydropower dam in Laos, intended to sell 995 MW of power to Thailand, has been widely accused of causing massive deforestation and the relocation of some 5,000 inhabitants from the Nakai Plateau. In 1995, the government of Laos asked the World Bank for support in the form of a risk guarantee to cover the US$100 million investment cost. Because the Bank would not extend this support until, *inter alia*, the project met with concurrence from social and environmental groups (Ganjanakhundee 2002: 3A), it was claimed by the Lao government that the affected community had accepted a resettlement plan elaborated for it (*Bangkok Post*, 22 January 1999, p. 5).

According to Joern Kristensen, former CEO of the MRC (Kristensen 2002: 4), most governments in the Mekong River basin find no particular difficulty in principle in agreeing to public participation but are faced with practical problems in adopting it. He maintains that the problem of how to involve stakeholders effectively in environmental decisions and the planning process has confronted the government not only in Thailand but also in other countries in the basin. This is compounded, so he argues, by potential conflicts of interest between communities at different levels, which may be local, national and international. Local communities may be seen to oppose projects planted in their midst that are in the national interest or they may support projects that are not in the national interest. They may thus put local or sectional interests above the national interest or even serve as proxies for unidentified vested interests. Unfortunately, all too frequently such communities do not speak with one voice, and it is up to the powers that be to decide which segment to listen to and take seriously. Again, concerned outsiders may go out of their way to support or oppose such projects, while the majority may remain conveniently silent.

Transboundary implications are even more intractable than domestic ones. Decisions taken within one country may well spill over into neighbouring countries. Thus, a country taking measures to protect its own forest resources may exacerbate illegal or unsustainable forestry activities in regional neighbours (Kristensen 2002: 4), as has been noted in the case of the impetus given to illegal logging in Myanmar, Laos and Cambodia by Thailand's draconian control measures. Moreover, the prevalence of illegal logging in Indonesia could be viewed partly as a product of the execution by certain Mekong River basin countries of forest protection policies that generate fresh demand for and escalating prices of regional timber.

Although it is true that any large water resources development project in the Upper Mekong could adversely affect millions of people in downstream countries, Kristensen argues that it requires much effort of the imagination to see how to involve the masses to be affected downstream in decision-making upstream. In the particular context of the Lower Mekong basin, problems of public participation have been compounded by poverty and the presence of many countries with differing national interests. Kristensen finds that the poor have limited access to the media and many have low levels of literacy and lack the skills and confidence to participate readily in public debate. He finds it is difficult to disseminate information effectively and it is equally difficult to secure responses to any proposed initiative, especially in countries such as Cambodia where civil society has been seriously disrupted by warfare and where the basic infrastructure is being rebuilt.

It is possible that Kristensen has exaggerated the practical difficulty of organizing public involvement in decision-making on a transboundary basis and has entirely overlooked the role of transboundary communication and transaction costs, which do limit the effectiveness of environmentalists' intervention (Benvenisti 2001: 117–118), not to mention the prevailing lack of political will on the part of governments in the Mekong River basin. Furthermore, there is of course no obligation on the part of any Mekong basin state to allow the public in other basin states to participate in its water resources development activities.

If *ex ante* public involvement has been found to be impossible even in Thailand, where democratic institutions and civil society are better developed than in the rest of the Mekong River basin (Panwudhiyanont 2002: 32), those affected by public sector water resources projects have had to resort to *ex post* protests. These have been found useful, especially where changes in the environment brought about by imposed public sector projects are not entirely irreversible. Indeed, the utility of *ex post* protests may in the short run be ad hoc in nature but could in the long run tilt the *de jure* power balance in favour of *ex ante* public participation.

Admittedly, a *post hoc* or *ex post* process, be it in the form of ratification, review or protest, suffers from three serious handicaps that could adversely affect its effectiveness and credibility. First, unlike an *ex ante* process, an *ex post* process cannot ensure adequate public scrutiny of a government's behaviour, because the government-as-agent, enjoying as it does the relative secrecy of a transaction, may find it comparatively easy to pursue partisan, short-term goals at the expense of its larger constituency (Benvenisti 2001: 117). Secondly, when the process takes the form of a protest, it is essentially negative in nature and generally offers no viable alternative to a public sector project confronting stakeholders with a fait accompli. It is unfortunate that civil society is all too frequently seen essentially as being active outside formal state political institutions and usually opposed to them (Dryzek 1996: 47). Finally, participation can be seen as the "new tyranny" (Cooke and Kothari 2001). This is not only because the concept has discursively and instrumentally extended the possibilities and modalities of co-optation (Editorial 2002: 1) but also because civil society tends to take up an extreme position and accepts no compromise.

Despite such drawbacks, two instances in the Mekong River basin do point to the apparent effectiveness of protests against faits accomplis in the form of public sector projects, since they have been able to undo much that has been done.

In its pre-regulation and pristine state, every nook and cranny of the bed and the wetlands of the Mun, a major Mekong tributary in Thailand, served as ideal habitat for fish in the flood season (especially from May to June) when fish migrated upstream for spawning. Groups of inhabitants in the area were in the habit of trapping big fish weighing 8–10 kg each, while allowing the rest to go free. In fact, each wet season catch used to be so copious as to allow the trappers to distribute it among relatives and sell the leftovers or preserve them with locally mined salt for subsequent bartering for rice. Similarly, in the dry season (especially from November to December) "hibernating" fish would return from cracks in the riverbed and wetlands to the river mainstream in search of a safe haven, and this would permit another large-scale fishing expedition. Resources were also invested in the purchase of fishing gear, in deepening and widening cracks in the river bed and in excavating small streams linking the river and the wetlands. In fact, claims to "ancestral" rights to fish-trapping areas and possessory rights to manufactured structures were so generally recognized in an atmosphere of communal solidarity and give-and-take that they are known to have been bought and sold openly. In addition to fishing, rice farming could be practised on the banks of the Mun even in the dry season owing to the ubiquity of water, but was particularly favoured by the advent of rainfall from April onwards. Equally, dry season vegetable horticulture took place on both banks of the Mun, where all manner of insects and reptiles, which throve there, would later become food for fish in the wet season.

This delicate ecological balance is said to have been destroyed by the construction of two dams, Rasi Salai and Pak Mun, which between them form an informal cascade. Seven years of impoundment behind the Rasi Salai dam, one of the most controversial public sector projects, is said to have caused nothing but devastation, and the distribution of irrigation water, which was the chief benefit claimed for the dam, was not effective (Chuskul 2001: 15). Public protests led to the opening of its seven sluice gates in July 2000 to alleviate the negative environmental and social impact of impoundment and to allow a land rights survey and a stocktaking of the situation. This revealed the spectre of submerged wetlands (*pa tam*) filled with decomposed plants and paddy fields cluttered with debris. In the wet season of 2000, after the opening of the dam gates, people reported sightings of huge fish (70–80 kg each) and even the much larger giant catfish migrating upstream from the Mekong, as well as "spectacular" catches after seven years of interruption. In one case, the reporter of the sighting could only stand idly by and watch, since he no longer had the right equipment with which to catch the big fish. Witnessing the fishes's homecoming, the inhabitants of the area have high hopes of the return of the good old times and the restoration of the natural ecological balance.

In a similar vein, in response to popular pressure, the sluice gates of the Pak Mun dam, which is situated quite close to the Mun's confluence with the Mekong, were opened, substantially improving fishermen's catches (Chomchai 2005).

Conclusions

Although the inhabitants of the Mekong River basin have been among the staunchest believers in and practitioners of traditional norms of governance, which correspond more or less to modern-day governance, the demands of large-scale and complex development have placed them in an embarrassing, disorienting situation. Prevailing property rights regimes, fiscal systems and patterns of economic growth have conspired to impose a radical departure from such norms, and widespread environmental degradation has ensued. At the same time, the basin's inhabitants have been no less devout in their adherence to the traditional participatory principle. However, modern-day participatory practice and principle are admittedly very different and not susceptible of instantaneous implantation. To turn these into home-grown counterparts calls for long-term development of civil society, whose present state gives cause for cautious optimism as well as concern.

The editors of *Mekong Update & Dialogue* (Editorial 2002) found that civil society's participation in resources management in Thailand had increased substantially over the previous decade as a result of the demise of military rule and the growth of democratic institutions, although elsewhere in the Mekong River basin things were not so rosy. Nevertheless, international NGOs have started to play a role in these other countries and local NGOs and specialized research institutes are slowly emerging with supportive roles, although they all have to operate within the confines of state-imposed totalitarianism.

Even in Thailand, however, where the prospects for civil society's development appear to be more promising, the concept of civil society is still elusive. Nevertheless, a number of groups are emerging and hopes are high that they will constitute the core of civil society in the very long run.

First, in parallel with the ongoing reform of Thailand's representative politics, adherents have emerged of what might be called "people's politics", "direct democracy" or "participatory democracy" (Rojanaphruk and Tangwisutijit 2000: A1–A2). One after another, grassroots community groups have risen in discontent against government projects, challenging the "classic" development model that appears to put sectional interests above the long-term well-being of the people as a whole. More and more people are unwilling to abandon their political rights the day after they cast their vote. Instead, they are to be found monitoring, criticizing and even intervening, if necessary, whenever politicians and bureaucrats are seen to abuse their power. What is novel in this is not just the unanimous sense of alienation from representative or electoral politics but a common conviction that active and direct democracy – as opposed to passive and unquestioning reception of government initiatives – is called for. This conviction is becoming more widespread and, once entrenched, is expected definitively to redefine the face of Thai politics and society. Since 2000, four separate but connected groups have emerged as major forces to be reckoned with in the arena of "people's politics". These are an alliance between various grassroots organizations and developmental and environmental NGOs; a coalition of middle-class civic groups; academics taken as a body; and a cluster of independent entities established under the constitution of 1997.

Secondly, as a society Thailand is quite open and closely scrutinized by relatively free mass media, which have played a critical watchdog role as part of civic organizations and grassroots groups. Such organizations and groups have become a formidable force to ensure that political leaders put the public interest before anything else (Chongkittavorn 2001: A4). Of course, media scrutiny is heavily focused on exposing scandals and corruption on the part of politicians and bureaucrats. Admittedly, without strong democratic institutional support, an emerging democracy such as Thailand's is fragile and extremely prone to manipulation by previous power brokers and various special interest groups. Although daily media exposure is intended to keep the government on an even keel, it could also gradually erode the popularity and legitimacy of elected leaders by appearing to put them on trial. The media themselves have to be transparent: they must assure the public that they have no axe to grind.

Whether and to what extent all such extra-parliamentary channels can be relied upon to take the lead and bring pressure to bear in favour of public participation on a more or less permanent basis remains to be seen, because things appear to be in a state of flux. The general public are baffled by street protests, which are daily occurrences especially in Bangkok, and the series of scandals highlighted by the mass media. In particular, the credibility of several groups, no matter how vociferous or articulate they may appear to be, has been put on probation by the silent majority. These groups invariably purport to act in a paternal role on behalf of those with grievances to air, although it is not easy to justify their competing claims to legitimacy or to ascertain to which constituency they hold themselves accountable. The onus is thus on them to prove their sincerity, impartiality and disinterestedness, since sometimes they bring pressure to bear to justify abuse of power by those in power; nor is it inconceivable that they could militate in favour of those wishing to regain power. The silent majority need to be reassured that they are not being made use of as if they were pawns in a political game of chess.

In support of the development of civil society, the MRC has not only invited civil society representatives to attend sessions of its Joint Committee and Council as observers but also incorporated the development of public participation as a component of all its core programmes, with particular emphasis on promoting participation at the sub-basin and local levels. In addition, it has provided assistance to agencies of member governments to develop their capacity to institute effective public participation activities (Kristensen 2002: 4).

Internationally, the MRC is being assisted by the Murray-Darling Basin Commission (MDBC) in Australia to develop its own public participation strategy for the Mekong River basin. Through joint workshops, study tours and training programmes, the MDBC model is being scrutinized and relevant approaches are being adapted. It is also intended to learn from some of the mistakes made in the Australian context (Kemp 2002: 4).

Even during the tenure of the Mekong Committee, the immediate predecessor to the MRC, modest beginnings were made with public involvement in project identification and development. It was here that the national Mekong committees played a strategic liaison role between the stakeholders in each member country and the Mekong Committee. This should continue to be a crucial role of the national Mekong committees operating under the MRC.

The crux of the matter is whether the government of a basin state voluntarily accepts the principle of public participation, particularly on a transboundary basis. Although the MRC has successfully brokered an agreement among the lower riparian countries on preliminary procedures for notification and prior consultation, it remains to be seen whether this will filter through. Without such an agreement, civil society would have to bring other pressure to bear, for instance through donor agencies such as the World Bank and the Asian Development Bank requiring public inputs before considering a loan, as happened in the case of the Nam Theun II hydroelectric dam in Laos. The MRC does not, however, cover China, which, like Myanmar, is no more than a "dialogue partner" to the MRC. Moreover, China has not applied to international donor agencies for financing for its Lancang cascade programme. Circumstances being what they are, civil society will have to be particularly resourceful to force public inputs on the Chinese.

REFERENCES

Bandyopadhyay, J. and V. Shiva (1989) "Development, Poverty and the Growth of the Green Movement in India", *The Ecologist* 19.

Bell, D. (2003) "The Making and Unmaking of Boundaries: A Contemporary Confucian Perspective", in A. Buchanan and M. Moore (eds), *States, Nations and Borders. The Ethics of Making Boundaries*. Cambridge: Cambridge University Press.

Benvenisti, E. (2001) "Domestic Politics and International Resources: What Role for International Law?" in M. Byers (ed.), *The Role of Law in International Politics*. Oxford: Oxford University Press.

Berg, H. van den (2001) *Economic Growth and Development – An Analysis of Our Greatest Economic Achievements and Our Most Exciting Challenges*. Irwin, IL: McGraw-Hill.

Bromley, D. W. (1991) *Environment and Economy: Property Rights and Public Policy*. Oxford: Basil Blackwell.

Can, Le Thac, Do Hong Phan and Le Quy An (2001) "Environmental Governance in Vietnam in a Regional Context", in Resources Policy Support Initiative, *Mekong Regional Environmental Governance: Perspectives on Opportunities and Challenges*. Chiang Mai, Thailand: REPSI.

Chantawong, M. (2002) "Civil Society Participation in River Basin Management: A New Blueprint?" *Mekong Update & Dialogue* (Sydney) 5(2).

Chambers, R. (1983) *Rural Development: Putting the Last First*. London: Longman.

——— (1994) "Participatory Rural Appraisal: Challenges, Potentials and Paradigm Shift", *World Development* 22(10).

Chomchai, P. (2005) "Public Participation in Watershed Management in Theory and Practice: A Mekong River Basin Perspective", in C. Bruch, L. Jansky, M. Nakayama and K. A. Salewicz (eds), *Public Participation in the Governance of International Freshwater Resources*, pp. 139–155. Tokyo: United Nations University Press.

Chongkittavorn, K. (2001) *Nation* (Bangkok), 12 February.

Chuskul, S. (2001) "Opening of Rasi Salai Dam – Return of Communal Life, Man, Fish and Wetlands", *Weekend Matichon* 21(1064), 8 January (in Thai).

Cooke, B. and U. Kothari, eds (2001) *Participation: The New Tyranny?* London: Zed Books.

Dryzek, J. S. (1996) *Democracy in Capitalist Times: Ideals, Limits, Struggles*. Oxford: Oxford University Press.
Editorial (2002) "Civil Society and River Basin Development", *Mekong Update & Dialogue* (Sydney) 5(2).
Environmental Research Institute (2001) "Public Participation in Development Projects in Lao PDR", in Resources Policy Support Initiative, *Mekong Regional Environmental Governance: Perspectives on Opportunities and Challenges*. Chiang Mai, Thailand: REPSI.
Ekachai, S. (2003) "Desert's Hidden Wealth", *Bangkok Post*, Outlook section, 26 March; available at ⟨http://pladaek.media.osaka-cu.ac.jp/limited/econetvis/⟩.
FAO [Food and Agriculture Organization of the United Nations] (2000) *Forestry Internet Site – Country Profiles*. Rome: FAO.
Ganjanakhundee, S. (2002) "Laos Signs $2-bn Deal for Nam Theun II Dam", *Nation* (Bangkok), 4 October.
Goodland, R. and H. E. Daly (1992) "Ten Reasons Why Northern Income Growth Is Not the Solution to Southern Poverty", in R. Goodland et al. (eds), *Population, Technology and Lifestyle: The Transition to Sustainability*. Washington, DC: Island Press.
Hanley, N., J. F. Shorgen and B. White (2001) *Introduction to Environmental Economics*. Oxford: Oxford University Press.
Hayes, D. (2001) *Japan, the Toothless Tiger*. Tokyo: Tuttle Publishing.
Higham, C. (2001) *The Civilization of Angkor*. London: Weidenfeld & Nicolson.
Hinton, P. (2000) "Where Nothing Is as It Seems: Between Southest [sic] China and Mainland Southeast Asia in the 'Post-Socialist' Era", in G. Evans, C. Hutton and Kuah Khun Eng (eds), *Where China Meets Southeast Asia: Social and Cultural Change in the Border Regions*. Bangkok: White Lotus.
Homer-Dixon, T. F., T. Boutwell, H. Jeffrey and G. Rathjens (1993) "Environmental Change and Violent Conflict", *Scientific American* 268.
Hourn, K. K. (2001) "The Impact of Regional Integration on the Governance Processes in Cambodia: The Environmental Perspective", in Resources Policy Support Initiative, *Mekong Regional Environmental Governance: Perspectives on Opportunities and Challenges*. Chiang Mai, Thailand: REPSI.
Hussen, A. M. (2000) *Principles of Environmental Economics – Economics, Ecology and Public Policy*. London and New York: Routledge.
Kemp, S. (2002) "Reflections on the Murray-Darling–Mekong Liaison", *Mekong Update & Dialogue* (Sydney) 5(2).
Kongkrut, A. (2003) "Rising Waters, Deeper Problems", *Bangkok Post*, 10 March.
Korten, D. C. (1991) "International Assistance: A Problem Posing as a Solution", *Development* 3(4).
Kristensen, J. (2002) "Civil Society and River Basin Development", *Mekong Update & Dialogue* (Sydney) 5(2).
Kungsawanich, U. (2001) "Out of the Woods", *Bangkok Post*, Outlook section, 10 February.
MRC Secretariat (1999) *Public Participation in the MRC*. Phnom Penh, Cambodia: Mekong River Commission.

Norgaard, R. B. (1994) *Development Betrayed: The End of Progress and Coevolutionary Revisioning of the Future.* London: Routledge.
Ojendal, J. and E. Torell (1997) *The Mighty Mekong Mystery.* Stockholm: SIDA.
O'Riordan, T., ed. (1997) *Ecotaxation.* London: Earthscan.
Ostrom, E. (1990) *Governing the Commons.* Cambridge: Cambridge University Press.
Panwudhiyanont, W. (2002) "Impoundment of the Pakmun Dam Has Brought Untold Misery to Mekong Inhabitants", *Sarakadee Magazine* (Bangkok) (in Thai).
Roach, J. (2003) "Saving the Giant Catfish: Mekong Giant May Face Journey's End", *Bangkok Post,* 10 June.
Rojanaphruk, P. and N. Tangwisutijit (2000) *Nation* (Bangkok), 31 December.
Ruangdit, P. and C. Theparat (2003) "Think-Tank Says B 70 bn Gone Down the Drain", *Bangkok Post,* 25 March.
Rudel, T. K. (1989) "Population, Development, and Tropical Deforestation", *Rural Sociology* 54(3).
Sacy, A. S. de (1999) *L'Asie du Sud-Est: L'unification a l'épreuve.* Paris: Libraire Vuibert.
Sakboon, M. (2002) "Putting the People First", *Nation* (Bangkok), 18 December.
Sluiter, L. (1992) *The Mekong Currency.* Bangkok: Project for Ecological Recovery TERRA.
Tangwisutijit, N. (2003) "How Will the Waters Flow?", *Nation* (Bangkok), 14 June.
TERRA (2002) "Declaration by the Local Communities of the River Basins in Thailand", Ubol Ratchatani, Thailand.
Tietenberg, T. (2003) *Environmental and Natural Resource Economics,* 6th edn. Boston: Addison Wesley.
Ting, Zuo (2001) "Cases of Local Transboundary Environmental Management in Border Areas of the Mekong Watershed in Yunnan, China", in Resources Policy Support Initiative, *Mekong Regional Environmental Governance: Perspectives on Opportunities and Challenges.* Chiang Mai, Thailand: REPSI.
Toope, S. J. (2001) "Emerging Patterns of Governance and International Law", in M. Byers (ed.), *The Role of Law in International Politics.* Oxford: Oxford University Press.
Turner, K., D. Pearce Kerry and I. Bateman (1993) *Environmental Economics: An Elementary Introduction.* Baltimore, MD: Johns Hopkins University Press.
Turton, A. (2000) "Introduction to Civility and Savagery", in Andrew Turton, *Social Identity in Tai States.* Richmond, UK: Curzon Press.
Wangvipula, R. (2003) "People to Have More Say in Water Use", *Bangkok Post,* 12 March.
Wong, E. (1999) *The Pocket Book of Tao.* Boston: Shambhala Publications.
Wongruang, P. (2002) "Mekong Fisherman Left High and Dry", *Nation* (Bangkok), 16 October.

Index

Access to justice
  public participation in TEIA, and 74–75
Africa
  TEIA in 58–59
African Union 59
Asia
  TEIA in 57–58

Bermejo basin
  GEF participation in 166–168
Border Environment Cooperation
    Commission 60–61
Botswana
  managing international river basin. *see*
    Okavango River

Danube-Black Sea basin
  GEF participation in 168–170
Data smog 84
Decision support systems 133–154
  advances in 142
  complexity of decision problems 133
  components 139
  decision problems in water resources
    management 134–138
    "best acceptable" solution 134–135
    complexity paradigm 137–138
    components of decision making
      process, diagram 134
    environmental processes 136–137
    operational decisions 135–136
    optimization techniques 137
    stakeholders 137
    technical developments 136
    traditional concepts, and 136
    two categories of 135
  examples of 141–142
  history of 138–139
  main building blocks, diagram 139
  mathematical models 140
  meaning 138
  prototype implementation of Web-based
    142–148
    case-study system 144
    characteristics 143–144
    consequences of selected policy
      alternative 148
    decision variables 144–145
    design principles 143
    impact of policy alternative
      146–148
    Internet, benefits of 142–143
    qualitative qualification of decision
      variables 146
    ranges of decisions variables 146
    simulations 145
    target user 143
  role and functions 140

218 INDEX

Discrimination
   public participation in TEIA, and 71
Dispute resolution
   public participation in TEIA, and 74–75

E-governance. *see* Internet
E-inclusion. *see* Internet
East African Community (EAC) 58–59
East Asian seas
   GEF participation in 171–173
Email-based field data collection system 120–132
   application to water infrastructure information collection 127–129
   environmental assessment, for 120–132
      acquisition of spatial information 122
      advances in IT 122
      cumulative impact assessment 120
      information gathering, reasons for 121
      nature of environmental crisis 121–122
      standard procedure for data compilation 122–123
   potential uses of 129–131
   system description 123–125
      details of client side, diagram 124
      server-side details, diagram 125
      system overview, diagram 124
   system implementation 126–127
   system use 125–126
      input data form, diagram 126
Environmental impact assessment (EIA)
   meaning 54
   role of, 14–15
   transboundary. *see* Transboundary environmental impact assessment
Environmental Kuznets Curve (EKC) 195–196
Espoo Convention
   implementation of 65–67
   TEIA, and 62–64
EU Council Directives 61–62
Europe
   TEIA in 61–64

Geographical information systems (GIS)
   role of 15–16
Global Environment Facility
   international waters 160–161
   process tools used by 157–179
      Bermejo basins 166–168
      Danube-Black Sea basin 168–170

East Asian seas 171–173
evaluation 164–165
freshwater basin case studies 166–171
Lake Tanganyika basin 170–171
learning, importance of 174–175
lessons learned from transboundary projects 175–176
local demonstration projects 165
marine ecosystem case studies 171–174
monitoring 164–165
national inter-ministerial committees 162
Pacific small island developing states 173–174
participation challenges 159–160
participation, for 161–165
public involvement policy 163–164
San Juan basin 166–168
South China Sea 171–173
strategic action programmes 162–163
transboundary diagnostic analysis 162
transparency 159–160
Green Cross International Water for Peace intervention 40–47

Harmonization
   public participation in TEIA, and 71
Helsinki Convention
   TEIA, and 62

Indonesia
   West Sumatra
      water resources management in. *see* West Sumatra
Information fatigue syndrome 84
Information technology
   role of 15
Institutional issues
   need to focus on 13–14
International Joint Commission
   TEIA, and 60
International waters
   Global Environment Facility, and 160–161
Internet 83–97
   10 step ladder of participatory democracy 84
   e-democracy 84
   e-governance 91–94
      democratic deficit 93
      European studies 92–93

INDEX

survey of 91–92
e-government 83
inclusive governance, and 86–89
  access to broadband in England and Wales, map 88
  digital divide 87
  e-Europe 88–89
  four aspects of content problem 87–88
  layers of connectivity 87
  perception divide, meaning 86–87
promoting on-line public participation 83–97
public participation in environmental governance, and 89–91
  e-governance, definition 89
  five possible benefits of 89
  four key lessons of on-line public participation 90
  implications of widespread use 89
  local stakeholders 90
risk and environmental information and 84–86
  complex nature of environmental problems 85
  liquid modernity 86
web-based decision support systems. *see* Decision support systems

La Paz Agreement
  TEIA, and 60–61
Lake Tanganyika basin
  GEF participation in 170–171
Legislation
  Indonesia, in
    Water Resources Law, 28–31. *see also* Western Sumatra
Local demonstration projects 165
Local rights
  failure to respect
    Western Sumatra, in 26–28

Man Wan Dam 198–199
  impact on downstream areas 199
Marine ecosystems
  critical problems facing 157–158
  GEF involvement in. *see* Global Environment Facility
Mekong River
  history of region 184
  location of 180
  poverty surrounding 181

public participation and governance 180–216
  access to information 203
  benefits of 205–206
  cash-crop plantations 183
  common property ownership 188
  conservation of water resources 184
  disruption to natural river flow 189
  ecological pricing 192
  economic growth 186–196
  economic reorientation 197
  effective involvement of stakeholders 208
  emergence of modern-day public participation 202–211
  environmental degradation 186–196
  Environmental Kuznets Curve (EKC) 195–196
  expert knowledge 196
  fiscal policy 192
  flooding 200
  globalization, and 190
  greening of taxation 192–193
  human population expansion 193
  Integrated Natural Resources Conservation project 183
  international instruments on 205
  irrigation systems 182
  local economy 195
  Man Wan Dam 198–199
  Mekong giant catfish, threats to 190
  "money politics" 203
  poverty alleviation 194
  prevalence of public participation 185–186
  promoted private sector projects 194
  property rights 186
  public finance, and 186
  public policy 186–196
  Res nullius property 188
  sale of natural resources 191–192
  state property regimes 187
  Taoist reverence for nature, and 181
  tax reform 193
  top-down public sector water resources projects 196–202
  totalitarian regimes, and 207
  traditional participatory principle 181–186
  traditional paternalism in river basin development 197

220 INDEX

Mekong River (cont.)
   transboundary implications 208
   wild tigers 191
   Yali Falls 200
  statistics 180

Namibia
  managing international river basin. *see*
    Okavango River
National Heritage Institute
  Sharing Waters project 47
National inter-ministerial committees 162
North America
  TEIA in 59–61
North American Agreement on
       Environmental Cooperation
  TEIA, and 59–60

Okavango River
  location of 33
  managing international river basin 33–52
    appropriate methodology for public participation 35–48
    core principles for development of methodology 36
    Eastern National Water Carrier in Namibia, map 38
    four key aspects of problem 35
    Green Cross International Water for Peace Intervention 40–47
    Namibian pipeline plans 37–39
    Namibian reaction to Botswana's strategy 39–40
    National Heritage Institute Sharing Waters project 47
    Southern Okavango Integrated Water Development Project 36–37
    Universities Partnership for Transboundary Waters 47
    Water Ecosystems Resources in Regional Development project 48
    water scarcity in Southern Africa, and 33–35
    Woodrow Wilson Centre Project 47–48

Participation
  enhancing 3–18
Peer-to-peer learning 98–119
  Integrated Water Resources Management (IWRM) 98–99

Millenium Development Goals (MDGs) 98
public participation, and 100
  application of 111
  learning agenda 113–115
  reasons for 105–108
  tools and techniques 111–113
  who should foster activities 108–109
  within adaptive management process for international waters 109–110
  within context of international water management 102
Process tools
  GEF, used by. *see* Global Environment Facility
Property rights 186
Public involvement policy 163–164
Public participation
  applying to water management 5
  appropriate methodology for 35–36
  financial resources for meaningful 73–74
  improving water resources management with 4
  meaning 4
  measures of successful 12–13
  Mekong River basin perspective. *see* Mekong River
  on-line. *see* Internet
  "parallel" 73
  Rio Declaration on 4–5
  role of 11
  scope of involvement 12
  tradition of in Western Sumatra 22–24
  transboundary environmental impact assessment, and. *see* Transboundary environmental impact assessment
  transition from dependence to empowerment 14
Public sector
  water resources projects 196–202

Res nullius property 188
Rio Declaration
  public participation, and 4–5

San Juan basin
  GEF participation in 166–168
South China sea
  GEF participation in 171–173

Southern Africa Development Community 59
Southern Okavango Integrated Water Development Project 36–37
Strategic action programmes 162–163

Transboundary diagnostic analysis 162
Transboundary environmental impact assessment (TEIA) 53–80
  benefits of public participation 54
  development of public participation in 70–75
    access to justice 74–75
    dispute resolution 74–75
    financial resources 73–74
    formalizing practices 71–72
    harmonization 71
    non-discrimination 71
    non-state actors 72–73
    "parallel" public participation 73
    specifity and clarity of terms of agreement 70–71
  environmental impact assessment, meaning 54–55
  failure to involve public 56
  local knowledge 55–56
  meaning 54
  in practice 64–70
    implementation of Espoo Convention 65–67
    Mexico and US 68–69
    Upper Mekong Navigation Improvement Project 69–70
    Victoria Falls 67–68
  public involvement in EIA 55
  public participation, and 56–57
  public's role 54–56
  regional TEIA initiatives 57–64
    Africa, in 58–59
    African Union 59
    Asia, in 57–58
    Border Environment Cooperation Commission 60–61
    East African Community 58–59
    Espoo Convention 62–64
    EU Council Directives 61–62
    Europe, in 61–64
    Helsinki Convention 62
    International Joint Commission 60
    La Paz Agreement 60–61
    North America, in 59–61
    North American Agreement on Environmental Cooperation 59–60
    Southern Africa Development Community 59

Universities Partnership for Transboundary Waters 47
Upper MEkong Navigation Improvement Project 69–70

Victoria Falls
  TEIA, and 67–68

Water
  improvements in efficiency 3
  knowledge about management 3–4
  management
    applying public participation to 5
    conventional approaches 6–10
    geographical scale and focus 11–12
    institutional issues 13–14
    international approaches 8–10
    IT approaches 7–8
    without public participation 6–10
  scarcity 3
Water Ecosystems Resources in Regional Development project 48
Water resource managers
  examination of various approaches 16–17
Water resources management
  decision problems in 134–138
  West Sumatra, in. see West Sumatra
Watershed development projects
  capacity development 10
West Sumatra
  economy 22
  public participation in 22–24
    community development 23
    creation of benefits 23
    land development 24
    musyawarah (public consultation) 22
    prerequisite, as 24
    traditional values of Minangkabau community 22
    unequal distribution of benefits 23
  water resources management 21–32
    access to information 29
    conflict among stakeholders 29
    developing cooperation 26

West Sumatra (cont.)
   drinking water 25
   failure to respect local rights 26–28
   hydroelectric plants 25
   new Water Resources Law 28–31
   public obligations 29
   public participation in 26–28
   pumping groundwater 25–26
   state control of 28
   state-owned enterprises 30
   total water potential 24–25
   "tragedy of the commons" 21
Woodrow Wilson Center 47–48

Yali Falls
   impact of 200